THE IRISH ARMY IN
1960–19(

C000254705

This book is dedicated to my parents:
Michael O'Donoghue (1909–1991)
and
Della Barton (1919–1990)

All the editor's royalties from sales of this book will be donated to the Kinshasa-based Street Girls of Congo programme. The SGC programme, which is run by the Congolese Catholic Community of America in Arizona, helps to take abandoned girls off the streets and place them in a safe environment.

THE IRISH ARMY
IN THE
CONGO
1960–1964

The Far Battalions

Editor

David O'Donoghue

IRISH ACADEMIC PRESS

DUBLIN • PORTLAND, OR

First published in 2006 by
IRISH ACADEMIC PRESS
44 Northumberland Road, Dublin 4, Ireland

and in the United States of America by
IRISH ACADEMIC PRESS
c/o ISBS, Suite 300
920 NE 58th Avenue
Portland, Oregon 97213-3786

Website: www.iap.ie

© David O'Donoghue 2005

British Library Cataloguing in Publication Data

A catalogue entry is available on request

ISBN 0-7165-2818-5 (cloth)
ISBN 0-7165-3319-7 (paper)

Library of Congress Cataloging-in-Publication Data

A catalog entry is available on request

All rights reserved. Without limiting the rights under copyright reserved alone, no part of this publication may be reproduced, stored in or introduced into a retrieval system, or transmitted, in any form or by any means (electronic, mechanical, photocopying, recording or otherwise), without the prior written permission of both the copyright owner and the above publisher of this book.

Printed by Creative Print & Design
Blaenau Gwent, Wales

Contents

List of Illustrations

1. Albertville 1960: left to right: Swedish Lt Stig von Bayer, Colonel Dick Bunworth, Cmdt P. D. Hogan, and Dr James 'The Badger' Burke
2. Michael O'Halloran , *Irish Press*; John Ross, Radio Éireann; Cathal O'Shannon, *Irish Times* in Elisabethville, Congo, August 1960
3. Rev. Daniel Dia-Mbwangi Diafwila
4. Lt Col Mortimer Buckley (retd.) examines a military map of the Congo
5. Belgian colonial administrator, Mr Emmanuel de Beer de Laer, at a Congo river crossing, 1951
6. J.J. Arthur Malu-Malu, a Congolese journalist working for the BBC World Service's African section
7. President Moïse Tshombe shakes hands with Belgian children in the mining town of Kolwezi, January 1962. Just behind Tshombe is Belgian lawyer, Mr Jacques de Jaer (centre, back)
8. Fleet Street reporter, Alan Bestic from Dublin, chats to armed Baluba tribesmen in the Congo, 1960
9. Comdt P. D. Hogan, third from left, flanked by Niemba massacre survivors Tom Kenny (left) and Joe Fitzpatrick (right)
10. Irish UN Troops search the railway line near Niemba after Balubas derail a train
11. Tom Kenny holds one of the arrows that wounded him at Niemba
12. Pte Tom Kenny is tended to by army medics following the Niemba ambush. Note two Baluba arrows hanging from his buttocks
13. Private Joseph Fitzpatrick at the Kamina Air Base, Congo, 1960
14. Stig von Bayer of the Swedish Army in UN uniform
15. Henriette Cardon-Sips (right) and her family take shelter behind sandbags in their Elisabethville home during a UN bombardment, December 1961
16. Pat Dunleavy (back row, first left) with Irish army comrades in Elisabethville, July 1961
17. Swedish Army interpreter Lars Fröberg (right) with Irish troops in Elisabethville after their release from captivity on 25 October 1961
18. Former Irish Army Chief of Staff Lt-General Louis Hogan
19. Dr Joe Laffan of the Army Medical Corps in Goma, Congo, August 1960
20. Officers of the 37th Battalion meet President de Valera before their departure for the Congo in May 1962. Fr Ronald Neville is in the front row, left
21. Fr Colum Swan was an Army Chaplain with the 2nd Infantry Group 1963–1964

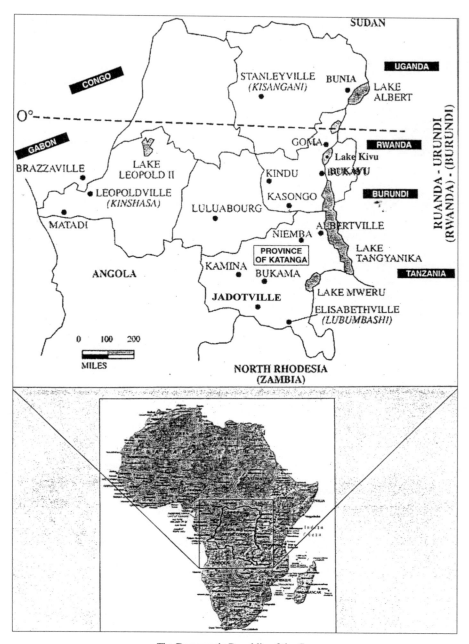

The Democratic Republic of the Congo

Acknowledgements

This book would never have come to fruition without the collective willingness of so many people to give freely of their time in recounting often painful memories of a fraught historical period. My overall idea was not to present a sterile, official history of the UN involvement in the Congo, but a living testimony from people who lived through the 1960–64 timeframe, when the newly independent Congo took its first faltering steps as a sovereign nation, and Irish troops embarked on their first major overseas mission. If readers are moved by the stirring individual stories from those difficult and dangerous times, then this book will have succeeded in its role as a record of intensely personal endeavour.

I am most grateful to all those who agreed to be interviewed about their personal experiences of the Congo, so that their stories could be included in this volume. They are too numerous to mention here, but their names appear at the beginning of each individual memoir.

To all those who agreed to submit written material for inclusion in this book, I owe a debt of thanks. They are: Colonel Ned Doyle (retd.), Dublin; Mr Cathal O'Shannon, Dublin; Rev Dr Daniel dia-Mbwangi Diafwila, Canada; Fr Colum Swan, Kildare; Fr Ronald Neville, Dublin; Mr Stig von Bayer, Sweden; and Mr Lars Fröberg, Sweden.

A number of others pointed me in the right direction while I was carrying out my researches on the Congo. They include: Professor Daniel Despas, Belgium; Mr Jean Leclercq, Afrika Museum, Tervuren; Ms Muadi Mukenge, Congolese Community of Southern California; Lt. Col. John P. Duggan (retd.), Dublin; the staff of the Military Archives, Dublin, including the director, Cmdt Victor Laing, Comdt Dermot O'Connor and Comdt Pat Brennan; the personnel of the Irish UN Veterans' Association, Dublin; the United Nations Training School Ireland, County Kildare; the administration section, Defence Forces Headquarters, Dublin; Mr Seán Ó Briain, whose expert knowledge of Irish was indispensable; Dr D'Lynn Waldron, California; Professor Mark Hull, St Louis University; Senator John Minihan, and my wife Claire for her help with French translations.

Finally, I wish to express my gratitude to Lisa Hyde of Irish Academic Press for all her help during the nail-biting and hair-pulling stages of preparing the book.

1. Albertville 1960: left to right: Swedish Lt Stig von Bayer, Colonel Dick Bunworth, Cmdt P. D. Hogan, and Dr James 'The Badger' Burke

Editor's Foreword

I remember being brought down as a child, in July 1960, by my mother and an aunt to see the soldiers parade past College Green in Dublin on their way to the distant Congo. I was only eight years old then and we lived on Baggot Street, which was a quiet residential area. I could not see much through the crowds, so I was hoisted up and could then see the serried ranks of troops marching past in all their splendour – caps and uniforms immaculate. I think a band was playing at the front. The sun was shining and everything seemed abuzz with excitement. I did not know what it was all about but I reckoned it must be important because I'd never been brought to the city centre before, apart from a trip to the top of Nelson's Pillar with my father and a journey on one of the last trams to Howth, circa 1959.

Looking back, those times seem like days of innocence, both for me and for the Irish Army whose members contributed so much to the UN peacekeeping effort in the newly independent Congo. It was the Irish Army's first major overseas mission since gaining independence from Britain in 1922.[1]

Back then, I had no way of knowing what path my life would take and nor had anyone else, I suppose. I could not have guessed that a quarter of a century later I would marry a woman who had been born in the Belgian Congo and whose family had to flee just as Irish troops were going into the troubled region. In their early twenties, my wife's parents had set off from Brussels on their great adventure into the unknown. Half a century later, the reflections of my father-in-law, Mr Emmanuel de Beer de Laer, on the mistakes of colonialism provide much food for thought.

Central to this book are the reminiscences of Irish troops who served in the Congo. In their own words, from humble private soldiers to the top brass, they spell out in detail what it was like to be there in the early 1960s, at the end of the colonial era and the fragile start of Congolese independence – another place, another time. Personal tales of suffering and tragedy are interwoven with comically absurd episodes, none of which detracts from the sacrifice of those who gave their lives in the service of the United Nations' mission to the Congo. In addition to those who paid the ultimate price for peacekeeping, there are many who will bear the physical and/or psychological scars for the rest of their days.

Overshadowing the army's involvement in the Congo from 1960 to 1964 – where over 5,000 Irish troops served – is the harrowing episode known as the Niemba ambush. Eight members of an eleven-strong patrol died in a hail of poisoned arrows, spears and gunfire on 8 November 1960, and the word 'Baluba' entered the Irish vocabulary as a pejorative term. Approximately twenty-five Balubas are thought to have died in return fire. Niemba still stirs bitter memories in army circles, particularly as the only two survivors have never been honoured.

Also controversial to this day is the so-called Jadotville incident, when 155

Irish troops were held prisoner by Katangese forces for over five weeks in September/October 1961. Jadotville veterans have been seeking recognition for decades. The Irish Defence Minister recently accepted that the troops involved had 'acted apppropriately' and stated that their contribution 'should be given appropriate recognition'. Also, in September 1961, three Irish soldiers died during Operation Morthar to restore order and UN control in Elisabethville, capital of the breakaway mineral-rich province of Katanga. Another four Irish soldiers died during the 'Battle for the Tunnel' in December 1961 when UN troops fought for control of Elisabethville.

This book tries to shed fresh light on what happened through the first-hand accounts of those who were there. In addition to the personal recollections of Irish UN veterans, the book includes reminiscences from: Dr Conor Cruise O'Brien, the UN Secretary-General's special representative in Katanga; Irish journalists who covered the Congo story; Irish Army chaplains; an officer with the Army Medical Corps; two retired Belgians, one an administrator in the Eastern Province, the other a prosecutor in Katanga during the secession period; a military historian; two Swedish army translators who worked for the Irish Army in the wake of the Niemba ambush and at Jadotville; a Congolese journalist and a Congolese clergyman.

By including Irish, Belgian and Congolese viewpoints, this book strives to achieve some sort of balance in recognising that Congolese independence affected many people's lives in various countries. To this day, Irish UN veterans meet regularly at their headquarters in Dublin to swap stories of those now distant events. One Irish veteran complained to me that army chiefs had sent weapons to one Congo destination while the ammunition was dispatched to another location almost 1,000 km away. Another veteran commented: 'The worst enemies we had in the Congo were our own officers.'

Meanwhile, on the first Tuesday of each month, Belgians who lived and worked in the Congo get together in Brussels for a mid-morning 'cocktail' to share their memories of those days. At one such gathering I attended, a retired Belgian deputy police chief in Katanga told me: 'If the Americans had been cut in on the deal, Katanga would have survived as an independent state'.

But the historic events following the hand-over of power by Belgium to an independent Republic of the Congo, on 30 June 1960, may be viewed differently by different groups. For example, in carrying out the background research for this book, I came across the Congolese Community of Southern California whose members, in 2003, launched the Congolese Genocide Memorial in Los Angeles. The memorial is 'to honour the 5.5 million Congolese that were killed in the invasion/civil war (1998–2003) that is still underway in the Democratic Republic of Congo, along with the 10 million Congolese that were murdered during our colonial period under King Léopold II of Belgium, and Lubas who were killed and displaced during the rule of the late president Mobutu Sese Seko'. The figures are staggering and place the events of 1960–1964 in context.

Our thoughts are for all those who have died in conflict in the Congo. Here in Ireland, naturally, we think of those veterans whose lives are forever linked to that watershed UN peacekeeping mission.

DAVID O'DONOGHUE
May 2005

Preface

Statistics, dates and official reports tell one story of any human activity. But the flow of events in which quick decisions have to be made, and the day-to-day effects of those decisions on individuals, are not conveyed by documents. The personalities of the decision-makers and of the troops who respond are rarely documented.

These intangible things are in people's minds; they blur or become distorted as memories fade. Ultimately they die with those who experienced them. Against this natural process we have eyewitness accounts. Not enough of these come from the soldiers and it is good to see many in David O'Donoghue's book. Memories, of course, are subjective. We occasionally see photographs of emplaning troops – now looking impossibly young – which remind us of time's ravages.

The frank views of M. de Beer de Laer, a former colonial administrator in the Belgian Congo, provide valuable and considered insights of a type we rarely hear.

It is forty-five years since our first troops went overseas to serve the United Nations and the cause of peace: things bigger than themselves and their own country. We had a very different Army in a very different world. The aim was to save lives rather than destroy them. Small countries like ours depend for survival on a peaceful world, where international treaties are respected and the rule of law is maintained. So, in a vital way, supporting the UN is a national interest.

The United Nations' peacekeeping operation (ONUC – Organisation des Nations Unies au Congo) was an early, stumbling step on the way to a new vision of international relations for the old ways had clearly failed.

After World War I, the demobbed soldiers looked back at the mud, the barbed wire and the scything machine-guns which had destroyed so many of their comrades. They said 'Never again' and 'Nie Wieder'. The politicians saw that the slaughter was utterly disproportionate to the causes. It seemed to have been an avoidable conflict.

Many felt that the automatic mobilisations and huge troop movements, ruled by railway time-tables, might have been halted had there been an international forum in which statesmen could have discussed matters calmly and rationally. And so, the League of Nations was set up.

But America, whose idea it was, did not join. The other Great Powers tended to ignore it if their interests were affected. The League had dedicated officials and did good work but the bigger states did not support it or submit to its judgements.

In 1945, after the horrors of World War II, the victors' delegates met in San Francisco to set up a 'League of Nations with teeth' in the parlance of the time.

For many people it was another attempt to break the old cycle of blood calling for blood – a deliberate plan to rule out future wars.

Think of those men – for almost all the decision-makers were men – politicians, diplomats and soldiers, over fifty years of age in 1945. They had lived through two world wars, the nullification of the League of Nations and the consequences: devastation, concentration camps, barbaric cruelty, disease, famine and the movement of voiceless refugee hordes.

Those graying men really wished to avoid a recurrence. The Charter of the United Nations reflects that. But the mind-sets and habits of hundreds of years of political, diplomatic and military education remained. National interest, as Mr Churchill had pointed out, was the best indicator of the actions of great nations. (He might have added 'ideology', often a prop for national interests in the twentieth century.) Because of the permanent members' vetoes, Security Council effectiveness depended on agreement. That agreement was already breaking down over Poland.

General Charles de Gaulle had made the first 'wind of change' speech in Africa in 1944, long before British Prime Minister, Harold Macmillan, made another one in 1959. The Belgian government, working to a much longer time scale, could not stand against the independence demands sweeping Africa.

Indeed, national interests and ideology did determine the attitudes and actions of the Great Powers to the Congo problems, when that huge state, almost as large as Europe, gained independence from Belgium in 1960. An attempt at secession by the mineral-rich province of Katanga was made. There was a breakdown in the discipline of the Congolese Army. The Congo's Prime Minister, Patrice Lumumba, sought UN assistance. And so, ONUC, the second UN peacekeeping force, was formed.

The term 'peacekeeping' is not in the UN Charter. It was an *ad hoc* invention to get Britain, France and Israel out of Egypt, after political miscalculation (and 600 dead Egyptian civilians) in 1956. It worked. Similarly, UNOGIL, the UN Observer Group in Lebanon, defused, or at least postponed, Lebanon's civil war in 1958. (The League of Nations had also run several successful peacekeeping operations between the world wars, but they are forgotten now.)

Everything was new in 1960. Dag Hammarskjöld, the great UN Secretary General of the 1950s, had formulated some principles and rules for the *modus operandi* with General Burns, the Canadian Commander of UNEF (United Nations Expedition Force). These could only be the beginnings of a peacekeeping doctrine, but they met the test of time as far as the 'consent' type of peacekeeping operations was concerned.

There was much to be done before dispatching the two Irish battalions. The soldiers were instructed in the new rules. Patience, negotiations and strict constraints on weapon usage were inculcated. The first two battalions requested by the UN were oversubscribed by volunteers, as were the subsequent ones.

There were vicissitudes and tragic losses. The *Daily Telegraph* led British news media in attacks on the force. We thought then that the British and French interest in Katanga's mineral wealth motivated the attacks. We know now that this was not the whole truth.

Professor Alan James's researches (see *Britain and the Congo Crisis, 1960–63*, Macmillan Press, 1996) have shown that Britain (in particular, Sir Alec Douglas Hume, the Foreign Minister) was hoping, even at that late stage, to salvage the concept of a Central African Federation consisting of the two Rhodesias and Nyasaland. An independent Katanga was needed as a buttress to the Federation's western flank. Professor James's book is essential reading in this regard.

Moise Tshombe, the properly elected president of the *province of Katanga*, had declared himself the president of an État du Katanga – a state recognised by nobody, but covertly supported by powerful foreign mining and political interests.

Whatever its merits or demerits, the Federation was a lost cause by the early 1960s. The black populations and many liberal Europeans saw it as continued white rule behind black puppet politicians. They may have been wrong, but the white personalities involved hardly inspired black confidence. The persistent efforts (covert and overt) to hive off Katanga cost the lives of UN soldiers and Katangese people - both civilians and hapless Gendarmerie – and of one of the greatest UN Secretaries General, Dag Hammarskjöld.

Lumumba was regarded as a tragic, intense figure with many talents, but gifted or cursed with a fatal oratorical facility that sometimes degenerated into dangerous invective. Tshombe, in his own way, was also a tragedy. He will be remembered in Afro-Asia as a quisling. He could, on occasion, show considerable charm and a sense of humour. He seemed to realise that the Congo, like much of Africa, still needed some outside help with education, technology and administration. The terms under which that help would be acceptable and workable were never really thought out. That applied to many places in Africa.

What of the UN effort? The revelations about President Mobutu and the fighting which spread from the Western Congo in the 1990s have led some of our soldiers to wonder if the UN effort was worth it all.

In retrospect, we can re-read with gratitude the balanced judgements of that most unusual British diplomat, the late and sadly missed Sir Anthony Parsons, former British Ambassador to the United Nations. In his memoirs, entitled *From Cold War to Hot Peace* (Michael Joseph, 1995), he judges that ONUC was a success:

> Against appalling odds, bombarded by criticism and invective, working in the uncharted territory of domestic conflict in a newly independent state at a time of maximal East/West rivalry, the UN operation maintained the integrity of the state... helped to restore law and order, mediated the end of the civil war and brought about the removal of disruptive European and other elements from the scene.

He goes on to say that, but for the UN operation:

> The Congo would have disintegrated, the infection of fragmentation would have spread throughout the continent, and the Soviet Union and the United States with their respective allies would have entered the fray... in a mounting number of mutually hostile, tribally based territories – in

short, a multiplicity of Angolas... When the eleven African and three Asian governments, plus Ireland and Sweden, committed infantry units to the Congo, they knew that they might have to fight rather than act only as diplomats in uniform. When the time came, they did not flinch and the bluff of the opposition was successfully called.

M. de Beer de Laer also says that 'without Katanga, the Congo would not be viable', although his honesty makes him add that he 'didn't think so in 1961'.

The Great Powers that criticised us have not done very well at peacekeeping themselves. Kosovo is half the size of Wales, but had 50,000 peacekeeping troops – twice the highest strength in the Congo, which is almost as big as Europe. Sarajevo, in Bosnia, was shelled almost daily for forty months – and we were told that nothing could be done and nothing should be tried, though two big powers had troops there.

There was some 'competitive post-colonialism' in the Congo – perhaps a case of 'look how much better our ex-colonies are than others'. Much of this was spin-doctoring and dubious reporting by journalists unable to get to trouble spots.

There were headlines about a mutiny in the Mali Battalion that never happened. The reports of a real mutiny in the Ghana Battalion were muted, although its British Commander was shot and the battalion was whisked out. (This happened in Léopoldville under the journalists' noses!)

We should not forget the excellent reporting, under great difficulties, of our own media people. Unused to the heat, loaded with cameras and recording equipment (not to mention spare shirts and socks), never sure about telex and telephone facilities or even food, they had as tough a time as the troops. Nevertheless they kept up with the moves and filed excellent reports and pictures.

Although the soldiers' perspectives were inevitably restricted, their accounts are often vivid and shrewd. They are well balanced by the officers' views. Lessons emerge on training, briefing and discipline – and the considerable differences between individual battalions.

The Congo should be seen in perspective. Nigeria was to be a model for Africa, but its civil war seems to have caused a million casualties. There have been several *coups d'état* there and in Ghana. No final figures are available for the India/Pakistan independence fighting. (Half a million deaths and 16 million refugees is Gerald Segal's estimate.) Algerian independence was also bloody.

The highest figure seen by this writer for Congo independence and its immediate aftermath is 50,000 deaths. This was largely due to the long-term, tight Belgian grip on firearms before independence. Killing was club, bicycle-chain or bow-and-arrow work, unless the army got involved.

This unusual compilation of memories enables us to re-live an honourable past effort, and an important time in our Army's development.

COLONEL NED DOYLE (retd)
October 2004

1

The Journalist's Tale

Cathal O'Shannon covered the Congo as a reporter for The Irish Times. Here he looks back over four decades at the events as he reported them for readers at home in Ireland, and places them in their historical context.

In the month of August 1798, one of General Humbert's officers, marching with his thousand men from Kilcommin Strand to Killala, commented on the fact that the peasant Irish came out in droves to welcome them 'in the name', as he said, 'of God and the Virgin Mary, bringing down blessings on us and offering us scapulars and other religious trappings'.

Just over 160 years later, the Irish once more rushed onto the streets, this time in Dublin, to cheer on an army, an Irish one, which was marching off to what they all hoped was glory. A few years ago on the Pat Kenny Show on television, this parade of the 33rd Battalion was described by Private Thomas Kenny, a survivor of the Niemba ambush, as follows: 'We marched along O'Connell Street and the people couldn't do enough for us. We were covered with holy medals and scapulars that they were pushing at us, throwing at us.'

Two occasions where an Irish public felt the same emotions at the sight of marching men. It was a time when the soldier was a popular figure, honoured by mere civilians who for a brief time could not do enough for him. So, not a lot, it may seem, changes. Civilians have always acted emotionally when military bands play and men march off to war. And it is as a civilian that I write to remind all of us of the emotionally charged atmosphere of 1960 – now, in these more cynical times, almost hard to imagine.

It is over forty years since men of the 32nd and 33rd Infantry Battalions set off for the Congo, the first armed Irish military formations requested by the United Nations to go abroad to help keep the peace – or, rather, to restore it – in a foreign field. And I was one of those civilians who, as a journalist, went out to the Congo with them.

In July 1960 when the call came from New York for Irish soldiers to join with other nations and fly out to the Congo, Irish newspapers were more concerned with other events than the rumblings of rebellion and threats of massacre in a far away land. In the central criminal court, Desmond Shanahan and Dr Paul Singer were being tried for various machinations of the great scandal of Shanahan's Stamp Auctions; seamen at the B&I Line were on strike and the docks were closing down; Irish political correspondents were tipping John Fitzgerald Kennedy to win in an election which was still half a year away; and the Irish Times was writing about Charlie Chaplin being on holiday in Kerry.

There was no television, apart from the rather fuzzy reception of the BBC signal along the east coast, and ads in the *Evening Mail* offered Renault Dauphines for sale at £374. Ireland had been admitted to the United Nations Organisation five years previously and, in 1958, Irish officers went as UN observers to the Middle East. By July 1960, however, more than observers were needed, and a perusal of Irish newspapers indicates a growing anxiety about what was happening in the Congo.

In June the African country had become independent of Belgium; in July the army mutinied and the Belgians sent in troops; Patrice Lumumba (Congo's Prime Minister) asked the UN to send in forces to keep the peace and restore order, and, incidentally, freedom. Ireland was asked to contribute men to this force. On the face of it, a simple request. But it obviously faced the Irish government with a dilemma. For a start, the various Defence Acts didn't cater for Irish troops serving overseas. They were – and they still are – *Defence* Forces to provide defence for this State. Soldiers who joined the army did so on the understanding, tacit or otherwise, that they would serve at home. In order to send them overseas on UN service these Acts would have to be amended, and so they were.

It was announced that Ireland was to send an infantry battalion to the Congo, and 600 volunteers were needed. Within a matter of days 3,000 men, more than a quarter of the army, had offered their services. Within days they were being formed into the 32nd Battalion, commanded by Lt. Col. Murt Buckley, and including such men as the 'Basher' O'Brien, Joe Adams, Commandant J.P. Laffin, Paddy Liddy, Donal Sweeney, and many other officers and men that I was to get to know well in the months that were to follow.

It's interesting to read the Dáil Debates for that period and see the anxieties that were concerning our legislators at the time. Paramount was: the need for an assurance that the troops would not be carelessly endangered; a concern that they should be officered by Irishmen; that their rations would be adequate; and their equipment suitable. Their moral welfare worried some people. The Independent TD, Jack McQuillan, suggested that the government 'in its exuberance and flamboyance' had not given enough consideration to the UN request. Joe Sheridan wanted the Irish troops to be known as the Casement Brigade because of Roger Casement's involvement with the Congo before the First World War; Noel Lemass asked that a contingent from the FCA (An Fórsa Cosanta Áitiúil – Local Defence Force) should go out with the troops; and Noel Browne wanted an assurance that Irish soldiers would not be used to buttress great financial institutions in the Congo, but that they should help return the country to its own people.

By the end of the month the officers and men of the 32nd Battalion assembled on the barrack square outside Collins Barracks in their thick bullswool uniforms, their peaked caps and leather leggings, and prepared to be flown off, via

Wheelus base in Tripoli, to what Conrad referred to as the Heart of Darkness, the centre of black Africa, the new Republic of the Congo. And the more you think of it, the more extraordinary it all now seems. Here was an army which grew out of our own civil war, with a minimum number of men, a good deal of obsolete equipment – or obsolescent, anyway – with no experience overseas, certainly no experience of tropical lands, new to the whole UN ideal, and inexperienced. That they set out with light hearts and great hopes is a bloody credit to them. They went willingly, almost joyfully, to a country of which very few of them had ever heard. Of its politics they knew little or nothing. They couldn't envisage its size or the complexities of its internal affairs. They had no language in common. They had no notion of what it might be like to work under UN command, or with the troops of other nations, many of whom were from former colonies now recently independent – Indians, Nigerians and Moroccans, as well as troops from Ethiopia, Sweden and Canada.

At this distance – and remember, it's over forty years ago and most of the youngsters who went out to the Congo then are, like myself, old men now – it seems an almost foolhardy venture. It says a lot for the professionalism of the army that they got a battalion of troops ready to fly out in a very brief period, some would say a too hurried period altogether. By modern standards, however, these soldiers were pitifully lightly armed. These were the days of the old No. 4, .303 Lee Enfield rifle and the Vickers machine gun which was trailed around in a sort of homemade pram with bicycle wheels. NCOs carried Gustavs and there was, I think, a sprinkling of Brandt mortars. Later on in the Congo there were to be some military products of what Conor Cruise O'Brien called 'the Ruhr of Ireland', Thompsons of Carlow – but by and large this was a frighteningly lightly armed force.

The Belgians and the Force Publique, the native military formations, had the more modern FN repeating rifles, heavy machine guns, artillery and air power, as well as properly defended firm bases and years of experience of colonial rule with military forces. This didn't prevent men of the 32nd beating the bejazus out of the Force Publique at a friendly shoot on the Goma rifle range. The Irish proved the superior marksmen over a whole series of competitions.

The United Nations was, even then, a bureaucratic organisation, deeply entrenched in protocol. Although the Irish Army had its own very effective public relations and press department, it was decided that no press officer would go to the Congo, so the Irish journalists seeking accreditation to the Irish troops had to be accredited to the UN. Our own army obviously wanted Irish newsmen with their troops, but the UN officials were quite unhelpful, even obstructive. Fair enough, we had to make our own way out to the Congo with the advance party, but the UN was unhelpful once we got there, apart from providing vague briefings on the general situation. By dint of begging lifts from Canadian and Swedish airmen, the tiny Irish press corps – all four of us [Cathal O'Shannon, *Irish Times*; Raymond Smith, *Irish Independent*; Michael O'Halloran, *Irish Press*; and John Ross, Radio Éireann] – got to Goma ahead of the Irish troops,

2. Michael O'Halloran , *Irish Press*; John Ross,
Radio Éireann; Cathal O'Shannon, *Irish Times* in
Elisabethville, Congo, August 1960

where we were promptly arrested with our Swedish army interpreter, our two
Irish officers and three NCOs, all armed with Gustavs but without a single
round of ammunition between them. We were held in loose captivity in a local
hotel until a bunch of Moroccan troops came in and disarmed the Force
Publique company which was holding us there.

Within a couple of days the first elements of the 32nd Battalion arrived in
the big US globemaster planes. What made headlines for the *Daily Express* and
UPI (United Press International news agency) who arrived at the same time,
was not the fact that these were the first to serve overseas, but the fact that with
the battalion came Fr Cyril Crean, the army's head chaplain and a great old war-
rior who had landed in Normandy 24 hours after D-Day in June 1944. With Fr
Crean off the globemaster came his portable altar, a veteran like himself of the
war in Europe.

Within hours the troops were deployed and allocated their quarters around
Goma. Communications with the Belgian army across the border in Kisengi in
Rwanda, with the local Force Publique in Kivu, and with civilians, black and
white, in the area, was hampered by the fact that only one officer with the battalion — a young second lieutenant – could speak French. No one spoke Swahili
or Lingala, so the UN had provided two Swedes, both former missionaries, who
spoke both French and Swahili. It seemed to me, however, that they were much
more anxious that the armed Force Publique be placated than obey the dictates
of the UN. The Force Publique for its part were more anxious that the Irish
troops feed and pay them (they hadn't been paid for weeks) than anything else.

After a fortnight or so, I found an Irishman, a planter who had fled the

Congo to Kampala in Uganda, who was willing to come back to the Congo and act as an interpreter with the Irish battalion. This was Mick Nolan, originally from West Cork, who had spent more than thirty years around the Ituri forest growing coffee, before he and other white men fled in panic when the Congolese army mutinied. Mick, who had been in the Irish army between 1925 and 1928 (or so he claimed) spoke French, Lingala and Swahili, and remained with the 32nd and 33rd Battalions for almost a year, ending up, I think, as a prisoner in Jadotville and Kolwezi the following year.

So, here was this Irish battalion, a thousand miles or more from the main UN HQ base in Léopoldville, left very much to its own devices, left to sort out its own problems in its own way. Its objective was not to disarm the Force Publique but to see that it performed its duties in an orderly fashion. The battalion was to restore peace and tranquility, to protect those who needed protection, and to keep communications open. Many of the Belgians who remained in the area, in places like Kindu and the other towns around the province, were opposed to independence for the Congo and wanted nothing to do with the United Nations – apart, that is, from protection by its soldiers when their lives or property were threatened.

Communications were appalling and must have been a nightmare for the staff of the battalion. Whatever radios they had brought with them had a maximum range of about forty miles, and the UN itself was unable at that stage to provide wireless equipment that could keep its HQ in touch with the troops in the field. Communication, therefore, had to be by air, or occasionally, by civilian wireless or telex. But since most of the internal services in the Congo had by that time gone for a Burton, there could be little dependence on these. As luck would have it, however, a police superintendent in Kampala, 500 miles away in Uganda, learned of the plight of the radio-less UN soldiers in Kivu. An Irishman himself, he not only had access to police radios but, as a ham radio enthusiast, was in communication with a priest in Dublin, a Fr Stone, who was also a radio ham. Between the two of them they managed to set up a link between GHQ in Dublin and the men of the 32nd. My recollection is that this policeman eventually sent a radio set to Goma which had a greater range than anything available through the UN.

Links of this sort were important because, as J.P. Duggan writes in his *A History of the Irish Army* (Gill and Macmillan Dublin, 1991, p.251), 'cafard, the result of members being removed from the family, and the effects of alcohol on Irish culture and society' soon set in. Post was slow in arriving and after a few weeks the excitement of being abroad faded and there were anxieties about what might be happening to families at home. When the post did at last arrive, the first Irish newspapers with our reports of the Congo came with them. These were carefully perused by Colonel Buckley – a man who had been somewhat apprehensive about what Irish journalists might be saying about his men and their actions. As I said, no army press officer travelled out to the Congo – I believe that was a mistake, incidentally, caused by the inbuilt suspicions of the civil servants in the Department of Defence and the Minister for External Affairs, Frank Aiken. Men like Mortimer Buckley would have had little truck with newspapermen during his distinguished career in the army, so he was naturally cautious in dealing with us. Having seen what we had reported in those first newspapers to arrive in the Congo, however, he was a lot more forthcoming – as, indeed, were his officers and men.

The 32nd Battalion was a lucky one. It never had to fire a shot in anger, though it might have had to threaten to. It patrolled thousands of miles – tens of thousands, I suppose – in its inadequate transport, the few Land Rovers it brought with it and locally bought or hired trucks and three-tonners. It placated dissidents, calmed down hot situations and brought refugees and others to places of safety. It was remarkable to see how rapidly these men, fresh from places like the Curragh and draughty barracks all over Ireland, came to terms with a new country and their new jobs. They succeeded by a mixture of charm and firmness, by fair dealing and by almost intuitive instinct.

These were, I think, the days of innocence: days when it was impossible to envisage this part of the Congo as it is now – torn by massacre and genocide, with more than a million Bahutu and Watutsi dead; days when it would have been impossible to foretell the losses our army was to endure in the coming months.

When I look back on those early days in the Congo a number of vignettes occur, most of them hilarious at the time. The local bishop who arrived unannounced and who sat for most of the afternoon swigging whiskey by the bottle. The black priest who arrived on a motorcycle with a lady riding pillion, and who was put up for the night; and the horror of the young soldier, who had to wake them in the morning, at finding the two of them in the same bed.

In his book, Lt. Colonel Duggan writes of the anxiety at the possibility of some of these young soldiers picking up social diseases from the compliant ladies of the Congo. Old Goma hands will remember the Café Mikeeno – café by name, knocking shop by nature. Mick Nolan, the Irish planter, came running to Commandant Martin O'Brien one night: 'You'll have to go down there, Martin', he said, 'there's black hoors, half naked, sitting on the laps of those lovely young soldiers and they're all jarred. They've spent all their money and one hoor is wearing a lad's scapular around her neck.' Amazing what can pass for currency in an emergency. The Basher O'Brien sorted out the owner of the Mikeeno in his own particular way. Commandant Laffan, who was the battalion's medical officer, was, of course, one of Ireland's leading veneriologists, and it didn't take him long to see that a supply of condoms was available for those who might feel inclined to dally with the native ladies of the night; a precaution which was welcomed and encouraged by that practical man, Father Crean.

By the time the 33rd Battalion was deployed trouble had erupted in Katanga. I don't want to go into the history of this period in any detail. Suffice it to say that Katanga, I suppose the richest province in the Congo, was to some extent the fiefdom of the giant Union Minière, a huge Belgian/British conglomerate supported by the Société Générale, the wealthy international financial organisation which was, to some extent, the creature of the British Conservative party. Katanga and its leader, Moïse Tshombe, threatened to secede from the rest of the Congo and become an independent entity. The United Nations' function was to see that this did not happen, and confrontation – political and military – was inevitable.

It is not for me to go into the tragic business of Niemba in November of 1960. By that time, all the Irish journalists had left the Congo, and it fell to us to cover the funerals of the dead in Dublin. It was at Niemba that innocence died for the Irish Army in the Congo and for the people at home. For the first time, Ireland saw the sons who had marched off so gaily putting their lives on the line thousands of miles away. The price they paid was for the freedom and peace of a country hardly known to them. Niemba made the word Baluba – the tribe involved in the Niemba ambush – a byword for evil, viciousness and treachery.

I will never forget the huge, awful outpouring of emotion, men and women in tears, as coffin followed coffin on the jeeps and gun carriages along O'Connell Street. The funeral marks an enduring, heart-rending scene for all who were there. The action at Niemba has been sifted over and over by military historians and others. General Carl von Horn, the Swedish Chief of Staff of UN troops in the Congo, questioned whether more battle-hardened troops would have met such a defeat, a comment ill received then and now.

Tens of thousands of Irish soldiers have perfected their trade in the service of the United Nations. It was in the Congo that they first learned that trade. Their monuments – both to those who served and survived, and to those who died – are there in the memories and minds of the people in whose lands they served, and in which they continue to serve with such distinction, ability, sacrifice and courage. The people of those lands, and you and I, owe them a debt of gratitude.

2

The Dream of an Independent Congo Died with Lumumba

The Reverend Daniel dia-Mbwangi Diafwila, PhD, a Congolese clergyman working in the Canadian capital, Ottawa, spells out his thoughts on the evolution of the Congo, which, he claims, was betrayed by the international powers.

Historically, Belgian rule came after the slave trade practices in Kongo, the ancient Kingdom of Congo – and the rest of Africa – including all the traditional kingdoms, which later formed the Belgian Congo and the colony of Rwanda Urundi. The process of colonization in the Congo was started by the exploratory enterprise of Henry Morton Stanley (1841–1904), the American newspaper reporter who had been sent to find the Rev. David Livingstone (1813–73). After successfully tracking down the Scottish missionary in 1871, Stanley was recruited by King Léopold II of Belgium, who wanted to extend his kingdom to Africa in order to strengthen his economic power.

While he was looking for Livingstone, Stanley had signed many blood covenants to win the loyalty of African leaders. Later on, he became the official representative of the Belgian King. During his last visit to Africa, in the Congo region, Stanley entered into blood covenants on many occasions with native chiefs, and even traded on such covenants before they were consummated. His description of the blood covenant rite is of objective and legal value, illustrating as it does the essential unity of such ceremonies the world over, although they may differ somewhat in minor detail.

When Stanley met with King Ngaliema of Kintambo – illegally renamed Léopoldville by the colonialists, without consulting local people, and now the capital city, Kinshasa – he was urgently pressed to made a personal blood covenant with him, and so put an end to all danger of conflict between them. To this Stanley assented, and the record of the transaction, on 8 April 1882, is given accordingly: 'Brotherhood with Ngaliema was performed. We crossed arms; and incision was made in each arm; some salt was placed in the wound, and then a mutual rubbing took place, while great fetish man of Kintambo pronounced an inconceivable number of curses on my head if ever I proved false' (H. M. Stanley, *The Congo and the Founding of its Free State* (New York: Harper & Broyhers, 1885) vol. 1, pp. 383–4).

Not to be outdone, Livingstone's former head man, Susi (then working with

3. Rev. Daniel Dia-Mbwangi Diafwila

Stanley), solicited the gods to visit unheard of, atrocious vengeances on
Ngaliema if he dared to make the slightest breach in the sacred brotherhood
which made him a Bula Matadi (breaker of rocks of great chief) 'one and indi-
visible for ever'. In June 1883, Stanley made the same type of covenant with
chief Mangobo of Irebu Bangala on the Upper Congo, chief Bakuti at Wangata,
and many other chiefs who strongly believed in his friendship and made him a
blood brother.

   As a result, the blood of a fair proportion of all the first families of
Equatorial Africa coursed through Stanley's veins. If ever there was an
American citizen who could appropriate unto himself the national motto 'E
Pluribus Unum', Stanley was such a man. But despite all the blood covenants
Stanley made with the chiefs and other representatives of our kingdoms, did he
ever think of asking the Belgian king, who had sent him to the Congo, to respect
these contracts? I do not think so because after Stanley's exploratory work,
Léopold II, the European king, transformed the Congo into his own territory. In
addition, he established his own companies to exploit the riches of the Congo
without seeking any collaboration from the indigenous tribal chiefs.

   While I regret having to use such strong language, King Léopold was nothing more than a thief. And the whole colonial enterprise, which came after the
so-called independent state of the Congo, existed to exploit the wealth of our
kingdoms, including the most important riches of which humanity can boast –
human beings themselves. Male workers were killed in the rubber plantations

just because they could not produce enough rubber for the 'boss'. The Congo made Belgium a great nation, with its diamonds, copper, uranium, cotton, lumber, ivory, rubber and other natural reserves.

Belgian rule in the Congo was inhuman. The colonialist did not consider an African as a person. He called his African men and women 'macaque' or 'monkey'. For him the African being was inferior to the European. This ethnocentric conception of the other and different human being was based on Levy Brulh's 'Ideological Anthropology' and the Bantu philosophy of Placide Tempels, a Catholic priest who thought he knew the African soul better, and affirmed the vital force as the traditional concept of being. He established a hierarchy of vital forces: at the top of this hierarchy was God, and after God the white vital force dominated the rest.

Belgian rule was humiliating for many workers, such as my father who for many years worked as a carpenter with the Congolese Office of Transport, building train carriages. His monthly salary was very low and when he retired he did not have enough money to survive. His sons and daughters became his social insurance. African workers could not defend their rights, including the right to earn good salaries. They did not have freedom of expression, action, movement or organisation. Even in the Catholic and Protestant churches, African priests and pastors were considered inferior to their European and American counterparts.

When I was growing up in the so-called Belgian Congo, every town comprised two very separate entities: one for the Europeans, which was called a town (*ville* in French); and the other for the Congolese, which was called *cité* in French (township). The indigenous population were referred to as basenji (savages) by the colonialists. The indigenous townships were for poor African people, while the town sector was for the rich, 'civilized' Europeans.

The greatest dream of many students of my generation was to become professors, medical doctors, lawyers or engineers in order to change their social status. The European colonial rulers created an intermediary social class they called '*évolués*' (developed people). An '*évolué*' was a man who could speak French very well, drink beer with white people, and work as an assistant for the Europeans. He had a bicycle for going to work and could wear a tie every day when going to his office. All other Africans who were undertaking manual jobs, such as cleaning homes, cooking meals in restaurants or building houses, were called 'Ngamba' or slaves – persons of no value.

Villages were totally separate from cities and no one could travel from city to village, or vice versa, without a special authorization from the district ruler, called a '*laissez-passer*'. The same permit system was used by Mobutu during his rule to control the movement of population between villages and cities. We were living in our own country as foreigners. We had a great fear of the Europeans who could easily use the police to arrest somebody, beat him or even kill him. I, personally, was unable to support this situation. I could not accept that a young white person could do whatever he wanted to my father just because under the colonial system native Congolese were considered to be naturally inferior to Europeans.

Despite the social discrimination of the system, colonial rule did many good things, like building churches, schools, hospitals and clinics, as well as providing training for various professional jobs, teaching hygiene principles and helping people to build modern houses in the cities. For example, following a colonialist's advice, my father built a house of 12 sq.m. for his family of seven children. I was very happy to live in it. I also enjoyed studying in the Belgian school, learning French, mathematics and other interesting subjects, which enabled me to progress in the sciences and various arts disciplines.

During the colonial period, a Congolese man called Simon Kimbangu started teaching, with the aid of the Bible, that black and white people were made equal by God and must work together for the progress of humanity. In 1921, in the rural city of Gombe Lutete, Lower Congo, Kimbangu made a prophetic declaration, stating that: 'One day we will get our independence. And the white will become black, and black will become white.' After bringing about a spiritual revival in many Bakongo villages, Simon Kimbangu was arrested and sent to jail with many other Congo prophets. He passed away in prison in 1951, but his prophecy came to pass nine years later.

One of the worst aspects of the colonial system was the total exclusion of African people from politics and urban administration. In 1959, the French President, General de Gaulle, delivered a speech in Brazzaville (capital of the French-controlled Congo) calling for the start of an emancipation process for the African nations. The Belgian colonialists were greatly surprised by this and would not agree to hand over control of the country's riches to the people of the Congo.

In response to the changing attitudes in France and elsewhere, some brilliant Belgian scholars created the ideological myth of premature independence. They affirmed the necessity of waiting for forty or even 100 years before granting freedom to the Congo because, they claimed, such a period was needed to train the future leaders of an independent country. They conveniently forgot that their ancestors from Belgium had met kings and chiefs everywhere they went in the Congo, and had agreed legal covenants with them, which the Belgian monarch did not respect. Thus, they were not able to build a healthy partnership with the African people in order to start developing a free and modern nation in the heart of Africa. They missed the train of history, were taken by surprise, and time began to run out for them.

African leaders were tired of colonialism and started to form political organisations to seek independence. The strongest and most important political grouping, which emerged in 1958, was the Alliance of BaKongo people (Abako). This organisation was an alliance of BaKongo leaders in Kinshasa, under the leadership of three key figures: Joseph Kasavubu, the first president of the Congo; Nlandu Nzeza; and Daniel Kanza, the father of Thomas Kanza, the first ambassador of Congo to the UN. Also influential was Sophie Kanza, the first female Congolese university graduate, who went on to become a great state leader. Many other Congolese representatives became members of this great association, Abako. Most of them had studied in Catholic or Protestant schools, were highly qualified and ready to take on the responsibilities associated with nationhood.

Apart from Abako, there were other strong political associations, including: the Bangala Alliance; the Baluba Alliance with Kalonji; the MNC (Mouvement National Congolais, or Congolese National Movement) of Patrice Lumumba, which was more national than ethnic; and the Conakat movement with Tshombe and Munongo in Katanga. The leaders of these key political movements went to Brussels for the round table negotiations in January 1960 to discuss the independence of the Congo. However, Lumumba was in prison at the time and, so, was unable to accompany his colleagues to Belgium.

At the same time, Mobutu was in Brussels pursuing a social studies course and training to be a journalist. Nobody recognized him as a potential Congolese political leader. When the round table conference began, all the Congolese delegates demanded that Patrice Lumumba must be present before any discussions on the future of their land could commence. The Belgian King, Baudouin, made a request for Lumumba to be freed and shortly afterwards he came to Brussels aboard a special Sabena flight.

When Lumumba arrived at the Brussels conference the proceedings gathered pace and became very interesting. Kasavubu and Lumumba sought total and unconditional independence for the Congo. At the conclusion of the round table conference it was decided that the Congo would become an independent republic on 30 June 1960. I was eight years old at the time, but I can remember the feeling I had when my father told me that we would once again become kings in our own country, would occupy all the Europeans' houses and start going to our fields by car. Wow! It was heaven on earth!

Independence Day, 30 June 1960, was a day of joy and hope. I can remember seeing our national flag being raised, and for the first time I saw my father crying with tears in his eyes. The time of freedom had come. All the Congolese dignitaries were gathered around President Joseph Kasavubu and the Prime Minister, Patrice Lumumba. (Both men had been democratically elected first as deputies and then to their respective governmental positions by the chamber of deputies and the senate, in accordance with the constitution agreed during the round table conference in Brussels.) They welcomed King Baudouin, who travelled to Léopoldville to take part in the official ceremonies to mark the birth of the Congo as a modern, sovereign and democratic state. It was the end of colonial slavery and the start of a new era of freedom and political responsibility.

The official ceremony granting independence to the new state took place in the Palais de la Nation in Kalina (known today as Gombe) – formerly a district reserved exclusively for Belgians. After speeches by King Baudouin and President Kasavubu, an incident (staged by the prime minister's advisers, including Mobutu, who were well prepared to stop from the inside the rise of this new star which had barely started to shine the light of political hope in the Congolese skies) marred the joy of national celebration for this first day of freedom. In fact, Emery Patrice Lumumba, who had been well prepared by his courtiers, rose to speak and in his address, unleashed a tirade against Belgium and the colonial system. While what he said was true, it embarrassed King Baudouin. His self-esteem damaged, the king decided to finish with this troublesome prime minister who had dared to show up the weaknesses and limits of the activities of the monarch's forebears in the Congo. From that day on, Lumumba's fate was sealed.

In reality, Lumumba fell into a trap set by his former tormentors who could never accept the fact that this former prisoner had become prime minister of the Congo and head of the biggest political party. The MNC united Congolese of all the former kingdoms in one nationalist, political structure with one ambition. That was to build a free and prosperous country in central Africa and to contribute to the liberation and development of the whole of Africa, with Africans from the north, south, east and west. In this respect, the leading figures were: Kwame N'krumah of Ghana, Sékou Touré of Guinea, Senghor of Sénégal, Ben Bela of Algeria, King Hassan II of Morocco, Habib Bourgiba of Tunisia, Nasser of Egypt, Emperor Haile Selassie of Ethiopia, Jomoh Kenyata of Kenya, Julius Nyere of Tanzania, Agustino Neto of Angola, Samora Machel of Mozambique, Kaunda of Northern Rhodesia and Abel Muzorewa of Southern Rhodesia.

All the elements for the liberation of the African continent were well established. Nasser, N'krumah and Sékou Touré were delighted when the Congo gained independence. The philosopher, Frantz Fanon, regarded Congolese independence as the trigger that would unleash a war for real economic and cultural independence in Africa – a continent that in ancient times had boasted the best organised kingdoms, including Egypt of the pharaohs where the Jewish patriarchs had sought exile and who later presided over the formation of the Israeli nation, having learned about political organisation and the human community from African priests in school. Jesus Christ did not escape that rule either. He also sought refuge in Egypt to escape the clutches of the little Idumean king, Herod, who wanted to kill Him.

Africa, the cradle of civilisation and freedom, began to rediscover its strength after Congolese independence. But a high-level plot, hatched in Belgium with CIA backing, would stifle all the hopes vested in the newly independent Congo. Mobutu, a bloodthirsty and ambitious man, was chosen to infiltrate Lumumba's inner circle with the aim of ousting the head of the army, thus more easily to eliminate Lumumba, Kasavubu, Tshombe and all those who envisioned the harmonious development of the country.

A few days after the independence celebrations, the president and prime minister began to criss-cross the country. I saw them with my own eyes in the town of Boma where I was living in 1960 with my parents. King Kasa, as we liked to call Kasavubu, accompanied Prime Minister Lumumba in a convertible car, waving to the crowds lining the streets of the Congo's first capital, Boma, where one can still find the Baobab tree where Henry Morton Stanley slept during his nineteenth century explorations of the Congo.

But the marriage between the president and prime minister of the new state would only last a few months. A mutiny by soldiers at Kokolo camp in Kinshasa drove the young prime minister to take a decision that would cost him his life and lead to the ruination of an entire nation for which a marvellous future had beckoned. Mobutu seduced Lumumba who appointed him as head of the Force Publique (national army). And the man from Lisala profited from this strategic appointment to halt the rise of his old boss and get rid of him.

In September 1960, Mobutu neutralised Kasavubu and Lumumba, and nominated a council of commissioners-general, which replaced the elected government. In January 1961, Lumumba's fate was sealed when he was sent to Elisabethville where his worst enemies were waiting, having been prepared to eliminate him since the first hours of independence. Lumumba was killed on 17 January 1961 in Elisabethville; his body was subjected to torture and mutilation before being dissolved in acid.

The Congo died along with Lumumba. Kasavubu was no more than a political wreck in the car park of history, waiting for the *enfant terrible* of Ubangui to dump him in the bin. He died at Boma, on 24 March 1968, two years and four months after the *coup d'état* carried out by Mobutu with the blessing of the international community led by Belgium and the United States. The UN's blue berets were complicit in this political assassination, paving the way for Mobutu's dictatorship, which in turn was replaced – with foreign backing – by Kabila's tyranny, on 17 May 1998.

In mid-January 2001, Laurent Désiré Kabila was assassinated in the presidential palace in Kinshasa. Since then, the country has been ruled by his son, Joseph Kabila.

3

To the Congo on a Wing and a Prayer

Lt Colonel Mortimer Buckley was the first Irish commander to enter the newly independent Congo, leading the 32nd Battalion there in July 1960. He looks back at the practical difficulties of initiating the army's first UN peacekeeping mission.

I held the rank of Lt Colonel and was Executive Officer of the Western Command when they decided to provide a battalion for the Congo. The Belgians were bailing out a year before they were expected to do so, leaving a vacuum. On 17 July 1960, I was in Galway on inspection and I was called back. I was told by phone to call off the inspection and to tell all the units to start collecting names of volunteers for six months' service in the Congo. All other activities stopped right away and I came back to Athlone. The following day I was in the office when the command O/C rang the bell. I went in to him and he asked me if I had a list of the officers of the Western Command who had volunteered for service. As it was, I had the list which I brought to him and he thanked me for it. After a while he rang again and he said: 'Your name is not on this list.' I said that I had not bothered including it. 'Do you not want to go?', he asked. Any Lt Colonel would go, so I replied: 'Of course. I have no problem with that. I would go if I was asked, but there is only one officer required. I know the Chief of Staff, General McKeown's form. He will go down the seniority list and go through it until he finds the man he wants.' The O/C replied: 'Oh, I suppose you're right. Okay,' and that was that. I went back into my office and at 12.50 p.m. he rang again. He looked at me and said: 'You know, you were right; the Chief of Staff did look down the list and he picked you.' 'Oh Jesus,' I said, 'I can't believe this.' I didn't expect it, I must say. 'Will you go?', he asked. 'Hold everything now,' I said, 'I'll have to see a doctor first.' I remember it well. 'Well go and see him', he said. I asked Charlie O'Neill, the command medical officer, to examine me and he said: 'You're for it, you're on the job.' He was the first one I told and I asked him not to tell anyone else. He ran his equipment over me and gave me the green light. I told the command O/C that I was on.

It was a very big responsibility but we didn't even know where the Congo was. No matter how good the 685 men were, we were going into a strange place, a different environment and climate. It was certainly new to me, but then I wasn't afraid of it. I was sent to Kivu province. Goma, where our headquarters were, is a lovely place. I had three rifle companies; they were fully organised and independent. I was in Goma with C-company. A-company was 120 miles away in Bukavu. They were the Dubs. B-company, from Cork, was in Kindu, 450 miles away. So, the question of going to their aid if they were in trouble was out. Nobody could be moved from the stations they were in. There were no means of moving them; we had no aircraft. We got aircraft from time to time to shift people around. There was only a small airstrip in Goma but it was able to take the American aircraft all right. There was an aircraft in Bukavu but it was in Rwanda, outside the boundary of our operations. It was only two miles away but we were under orders not to use it. A-company had to come up to Goma and go down by boat to Bukavu, which was 120 miles away. Bukavu was a lovely town. The troops were very happy there and had no great problems. Kindu, on the other hand, was on the shores of a big river, one of the tributaries of the River Congo. It was very low lying and very warm so I was always glad to get out of it. I had to go there a couple of times to see how they were getting on and to do the best I could to get them into accommodation, which we did.

At that time the Chief of Staff, General McKeown, was at home in Ireland. General von Horn was our UN commander in Léopoldville, 1,000 miles away from Goma. Communications were extremely difficult. General von Horn was Swedish with some Irish blood in his veins. I do not know how that came about but he often told me he had Irish blood.

We were airlifted out to the Congo from Baldonnel in American aircraft – big, powerful, four-engined propeller-driven C130s. They were most uncomfortable with canvas seats facing each other, with no elbow rests. We were not facing the line of travel, we were sideways. Sometimes it was cold and sometimes it was too hot, but we didn't care.

We had about four days in which to prepare for our departure. I was appointed on 17 July 1960 and the battalion was formed on 19 July. They met in the Curragh. One day all the officers were brought up to Áras an Uachtaráin to meet President de Valera. We had tea and coffee with him. He was talking to the Chief of Staff, General McKeown, and myself. Then, de Valera pulled me aside because he wanted to say a few things to me. 'I will give you this advice,' he said, 'Wherever you go, go in such strength that everybody will be aware of your presence and strength, and not of your weakness.' That is a very good solid principle of war, of course: to go with as many as you can and to make sure that you have enough. But when you are travelling long distances, you have to cut your cloth according to your measure. If you bring an extra jeep it means extra petrol, and bringing extra men means bringing extra food. Therefore, you have to cut down and balance the thing and take a calculated risk at times. De Valera's second piece of advice to me was: 'If you fire the first shot in any situation you've lost the battle. Don't ever forget that.' Of course, he was so right. I never forgot that and I saw it proved several times. I recall that occasion very well. I was an admirer of de Valera's. As President, he was Commander in Chief of the Army. That is what he told me and I was very impressed by it.

At 6.30 a.m. on 27 July we came up from the Curragh to Castle Yard,

Dublin. It was pouring rain but it cleared up around 11 o'clock and we marched through Dublin. It was the most fantastic experience I ever had. The battalion was going to the Congo but I thought there'd be no one out. I thought they might be out when we came back, if we were still alive. I was flabbergasted by the reception, marching in front of the battalion. There were fellows up on buildings cheering, and everything stopped. I was very impressed. From there we went to Baldonnell where we had a meal. We met hundreds of people and were addressed by the Archbishop of Dublin, Dr McQuaid. He gave us his blessing and told us that the Congolese people were very gentle and kind. He said they hadn't a notion of what it was they had got in this freedom, which they thought they had. He said they should be treated accordingly, with sympathy. I agreed with what the Archbishop said; he was a very learned man.

The first three planes left Baldonnell at 3.30 p.m. We landed in France at an American base at Evreux, just for half an hour to refuel. We took off and headed for the North African coast and landed in Libya at a big American air base. We were accommodated there for the night. The following morning we attended Mass there. Would you believe it? The American army chaplain there, Fr O'Brien, had been a neighbour of mine in Kerry. He lived about four miles from my house. I had my own two chaplains with me: Fr Cyril Crean and Fr Gregory Brophy, who was head chaplain in the Curragh.

We left Libya that evening and headed for Kano in Nigeria where we refuelled and had a meal. We only stayed there for about an hour. That was the first time the heat hit us. Our next stop was Léopoldville in the Congo where we arrived in the middle of the night. We were not allowed off the planes. They said that if we got off we would not be allowed on again, but they were just being stupid. They did not want to let us off in the middle of the night because the fellows might start wandering around and it would be impossible to collect them. We were very tired at that stage. From there we went straight to Goma and arrived at 7.30 in the morning with the sun shining clear. We were surrounded by the natives – black men, women and children. It was a bit frightening.

Comdt Joe Adams from Tullamore, who was my second in command, was there before me. He had been sent out with a small group to reconnoitre these areas in Goma, Bukavu and Kindu. He shouted up to me in the aeroplane: 'Have you got the band?' I said I had and he said: 'Get them ready, get them out.' So the pipe band marched out and the local people went mad with delight. We won the day without firing a shot, as the saying goes. After that we settled into three big school buildings. The school was closed at the time for holidays.

We had problems at the border but we were never in danger. I became very friendly with the local commissioner and assistant commissioner who were natives, black people. They dressed impeccably and spoke French fluently. We had several interpreters, about half a dozen. When the 33rd Battalion came out, about five weeks afterwards, we had to send some of our interpreters to them. We were always short of interpreters, so we were allowed to hire our own if we found any. We found a man called Mike Nolan, an Irishman, who understood

4. Lt Col Mortimer Buckley (retd.) examines a military map of the Congo

the native language because he had been there all his life. His job blew up when the Belgians left. He owned a ranch there, growing tea. We had several Swedish interpreters and a Russian lady, Miss Degalberg, who also owned a ranch on the shores of Kivu province. She was a fine woman. She knew eight or nine languages. There was another lady, a Belgian who lived in poor circumstances. She was an expert in the native languages. She wrote out a small vocabulary of words for us. She didn't like the natives and used to argue with them. If she was interpreting she would start doing the arguing, instead of me. We had to get rid of her in the end.

Occasionally, I would get a plane to go across to Kindu which was 450 miles away. We used old DC3 planes. On one occasion, five of us went across in a small twin-engined Cessna. It was a lovely little plane, just like getting into a car. We hit an electrical thunderstorm during the flight and spent an hour and a half in it. I can tell you that there is no more frightening thing in the world. Early on, I became reconciled to the fact that there was no hope we would survive it. I didn't say anything to the pilot because I presumed it was his job to know what to do – whether to turn and go back, or whatever. He kept going anyway. Half the time I didn't know whether we were going backwards or forwards. We got into a heavy cloud like a fog and we couldn't see out through the windscreen. We couldn't even see the propellers of the engine. The noise from the wind was unbearable, it was a tornado. The lightening was hopping off the cockpit windows. I had to close my eyes because I couldn't keep looking at it. At one stage I found myself standing up. I don't know what happened. I noticed that I had a hold of the pilot by the collar and he was standing up. The plane must have turned upside down because the joystick was up in the air and the pilot was standing trying to pull it down. There were no seat belts in the plane. After that, I put my hands in my trouser pockets so that I would not interfere, no matter what happened. I closed my eyes then and said every prayer I could think of, and said them several times. I could not see us getting out of it and I was worried that nobody would ever find us because we were flying over tropical forest. Eventually we flew out of the storm at the other side. Time was dragging on so much, I thought we must be half way across the Atlantic. I couldn't even work out in what direction we were going. I couldn't think straight. The pilot found a hole in the cloud and went down through it. We landed in Kindu with the sun shining. The flight had been a horrible experience, unbelievable. At one stage during the flight, I had my eyes closed and the noise abated. I was weightless and could not feel the seat under me. I thought I had left the plane and was out in the air. I had to put my hand on the seat to see if it was still there, and it was.

While in the Congo, we never came under fire but the nearest thing to a head-on clash happened in Goma when I was least expecting it. I thought I was the Lord Mayor of Goma and everyone was friendly there. I thought nothing could go wrong. On one occasion however – in September 1960 – all UN personnel got orders to hold all airports from 8 o'clock the following morning until a future date. After I received that order, over 100 Congolese soldiers arrived into Goma airfield. We always had a guard on it, but they didn't interfere with it, they were in a group at one side. I decided that instead of holding the airfield at 8 a.m., I would go out at 4 a.m. I selected six jeeps, six NCOs and about 25 men. I could only spare that number of men because Goma was being stripped all the time to feed another post that had opened in Kamina which was 300 miles away. Comdt Joe Allen was the O/C there. They were drawing troops away from me all the time. We went out at 4 a.m. and slowly drove the jeeps onto the airstrip, about 100 yards apart. The men got out of the jeeps and took up positions along the rocks so we were in possession of the airfield. As I thought, the Congolese soldiers who had arrived earlier were asleep in a huddle in a group. When daylight came they woke up to find that we were in possession of the airfield and they became obstreperous. They were threatening us with all kinds of things – what they would or wouldn't do – but I didn't think they would do them. Eventually, they stretched out in what is called a skirmish line, a line of men across the airstrip facing my 35 men, and started walking towards them standing up. They all had guns, of course. Our fellows started shouting at them: 'Stop, stop. Go back or we'll have to fire. Stop, or we'll have to fire.' We had a loud hailer and Mike Nolan the interpreter was there, but they wouldn't stop. They kept coming. I was there but I wasn't the commander. I had appointed an officer in command, Comdt Joe Dwyer, and I wasn't going to interfere with him. He did a wonderful job and his men never fired a shot. I thought they would panic but they didn't. The crowd was within two yards of a line and the officer had made up his mind that if they crossed it he would have to start firing because they were getting too close. They would jump in on top of us if we let them go any further. They stopped, stood there and eventually turned about and walked back. I was watching that and I thought there was nothing I can do to stop it, but it stopped itself. At first they thought we'd run or something. I don't know what they thought to be precise but they didn't fire and we didn't fire. It was very close that we'd fire the first shot, which I didn't want to do, and which, I think, they probably did want us to do. But it didn't happen, so that was that.

It was important for me not to fire the first shot because we were a peace-keeping force. Firing the first shot means you are a belligerent – you are law enforcing or peace enforcing, rather than peacekeeping. It is different now in many places where the UN operates because it is enforcement, they compel people. Our job was to maintain peace and to help in every way possible. We did an awful lot of patrols from Goma and other places. Luckily enough, I made a strict order that if we got into any situation where the road was blocked – which we did several times where fellows would block the road and say we couldn't pass – we would just wave goodbye, turn around and return to base. There was to be no confrontation or involvement. We would go back the following day. They had gone home, delighted with themselves, and the road was free for us.

The native soldiers always wanted to join with our troops to do guard duty with us. We could not allow that, of course. The language question alone would stop it. They also wanted to go on patrol with us. That strictly wasn't allowed either. One time, one of my staff said to me: 'Sure, what's the harm in bringing a couple of them with us when we are going on these patrols? Maybe they would help.' I thought about it for a couple of days and decided to try it. I thought that if it worked we would tell the General that we had done it, because there was a rule against it. It worked like an open door. Every time we planned to go on patrol in Goma, word would be sent to a barracks three miles away that we wanted two soldiers tomorrow. They would be down at 4 o'clock the following morning sitting on a bench outside our headquarters waiting to go. Of course, we used to feed them. They had their guns with them. We put them up on the jeep in front and when we went into any town these two guys would stand up and everyone was waving at us. You got the freedom of the whole place. In fairness, everyone saw that we were treating them the same as ourselves. I wrote to General von Horn to explain this to him and I recommended that UN patrols should be allowed to have these people with them whenever they were going out. He changed the rules accordingly. That was one of the best things we did.

I was invited into UN headquarters on 24 October 1960 which is known as UN day. It is a day for celebrating, playing football and bands playing. General von Horn invited me to Léopoldville and sent his plane for me that day. I brought a pipe band from Cork with me. They were a powerful unit. We enjoyed ourselves there for a couple of days but were glad to get out of it because it was a hot old spot. While there, I met Col. Justin McCarthy who was an Irish officer on General von Horn's staff. He was not attached to my 32nd battalion. He drove me around Léopoldville in his car and showed me all the places. I left the city on the 26 October at 4 o'clock in the morning. I eventually arrived by a circuitous route back in Goma late in the evening. The following morning a message was put on my table that Col McCarthy was dead. He had been killed in a terrible car accident. I couldn't believe it because I had been with him the day before, driving around the city in his car. General von Horn ordered me to report back to Léopoldville again and bring the band, 18 strong NCOs, a police officer and the head chaplain for the funeral. For diplomatic reasons, the funeral service took place at a native church rather than at the Jesuit church in Léopoldville, because it would look better. General Mobutu, who later became President, was there too. The coffin was later flown home to Ireland for burial.

I never saw any atrocities committed in the Congo. However, on one occasion I was invited back to Cork to B-company – my own battalion. Some of the NCOs there had photographs of atrocities and they showed them to Fr Crean and myself. I must say that I was shocked and I didn't like it. I don't know where they got the photographs. I don't know where they saw dead people. I never saw a corpse anywhere and before seeing those photographs I had never heard from anyone that they had seen any atrocities. But the photographs I saw were shocking. I would have burned them. I did not like the fellow who had them. There were atrocities there before we went out, we were told. We have to believe that there were Belgians killed and raped, but whether that is true or not I can't say. All I can say is that in any place I went to there was peace and quiet. I was lucky in Kivu province. It might have been different elsewhere, I could not answer that. But I didn't see any of that, or hear of it, during my stay there.

I don't know very much about the Niemba ambush. I was very near it the day it happened but I did not know about it. I was ordered down to Albertville, in my single-engined Otter plane, to meet Col Harry Byrne. I met Col Bunworth and his officers. We waited and waited but Col Byrne didn't turn up. It was a very wet evening. I came back to Goma and later that evening I received word that this had happened at Niemba. During the night I had a Jewish civilian who was running the high-powered transmitter which kept us in contact with Léopoldville and other stations. He got the names for me during the night. He picked them up on the air going across from Albertville to Léopoldville and on to Elisabethville. The following morning I had the names of the nine who died. At lunchtime the Chief of Staff was on the blower as we had our own transmitter.

It is a wonderful story of how we came by the transmitter. Cmdt. Deegan got in touch with a ham radio enthusiast in Uganda who transmitted all over the world. He said he would lend him a radio set while we were there so we could talk to a Fr Stone – another ham radio expert – who was a curate in Dublin. Comdt Deegan sought my permission to bring a patrol 200 miles to collect the radio. After some hesitation I allowed him to go. The man on the other side also had to travel 200 miles to meet him at the Ugandan border. He had made the radio set out of bits and pieces. We got it going and could talk to Fr Stone in Ireland. He was a wonderful man and made himself available at all times, although he never got any credit for doing so. General McKeown was in the priest's house, as he wanted to talk to me. He started asking me all about my battalion – how it was getting on, the weather, the clothing and food. I knew well that the tragedy had happened down the road the evening before. I kept asking myself: 'Does he know about this? Am I to be the one to tell him, if he doesn't?' As it turned out he did know. He never asked me whether I knew anything about Niemba. Eventually I said: 'Sir. Stop talking about my outfit where everything's grand, but there's trouble in the 33rd. There has been a terrible tragedy, and I am sorry if I am the first to have to tell you.' He said: 'No, we know about it but we don't know the names.' 'I have the names,' I said. I was able to give him the names and he was able to get people out to the houses [of the families of those who died]. General McKeown was very pleased to get the names of those who had died at Niemba.

We were very pleased to have that radio set. We used to get the results of football matches and hurling finals from Fr Stone before they were broadcast on the 6.30 news. We were deeply grateful to the man who gave us that radio set, even though we had to go a long way to get it. Eventually we had to hand it back to him when our tour of duty was over.

I do not know what went wrong at Niemba. I hate forming an opinion about it. They were caught out. It could happen to anyone, in my opinion, to be straight about it. They were probably a bit too friendly. I had difficulty all the time in warning the soldiers to watch themselves and to keep looking around. The Irish soldier is a very friendly guy when he is out anywhere. He is never aware or thinks about danger or anything like that. I suppose that is a good point, it is not bad. As regards Niemba, I cannot answer the question as to what went wrong there – much as I would like to – for the simple reason that I never read the official reports on it. I never had an opportunity of doing that. That's number one, and number two, you'd want to know the temperament of the place itself. I did hear that there was opposition to the movement of troops around there. In my wisdom – and I don't claim to be that clever – I would have stayed at home for a couple of days and let these people wear themselves out. That is being terribly wise now, and it is probably being unfair, but what happened to Lt Gleeson could have happened to any of my officers, once they were surrounded and they [the Balubas] started firing the arrows. These people were very wily people, but it didn't happen to any of my fellows – I had no casualties – and I was very lucky because it could have. I knew Col Bunworth, the commander of the 33rd battalion well; he had been a cadet with me for two years in the military college. We were great friends. We did not discuss Niemba afterwards. I couldn't do it with him. I know he submitted official reports to the Chief of Staff and to various authorities but I never saw them so I've no idea what is in them. I wouldn't think the reports were ever published. The thirty-year barrier is up now, but all those reports are in the archives in Dublin. I don't know whether they would allow me to see them, I doubt it. But I don't want to see them at this hour of my life. What do I want to know for? It happened and it was unfortunate.

After Niemba, both Col Bunworth and myself were called in [to the Department of Foreign Affairs, on 15 January 1961]. It was a very friendly meeting. According to Dr Conor Cruise O'Brien's book[2] we were both brought in separately and quizzed. That is an error. There was no question of being quizzed. We were invited to a meal by Frank Aiken, the Minister for Foreign Affairs, and Cruise O'Brien who was a senior man in the Department of Foreign Affairs. We hadn't a clue what they wanted to know or why they wanted to know it. We did not know that Cruise O'Brien was going to go out [to the Congo] in the next fortnight. There was no question of being taken in separately or anything like that. We were told to be at Iveagh House at 12 o'clock. I was at home on holidays. We got special leave. Bunworth was down in Cobh where he was from. We both arrived at Iveagh House at the appointed hour and met Frank Aiken and Cruise O'Brien. We spent no time at all there. We walked up to the Russell Hotel and were brought up to a room which they had prepared for us. We were given a meal and during the meal we were quizzed about life, what had happened and this, that and the other. Generally speaking, it was all palsey-walsey. I was in a laid back position, I had no problem with it. I didn't realise that Cruise O'Brien was looking for information to help him to go on foreign service, nor did he tell me that he was going. If he had told me I would have been more careful, possibly, and would have tried to be more helpful. I wasn't told anyway. After the meal we parted and everybody went home.

Editor's Note: According to Dr Conor Cruise O'Brien's account in *Memoir: My Life and Themes*, the meeting on 15 January 1961 took place in New York. Col Buckley is adamant it occured in Dublin. He disputes the following extract in Dr O'Brien's book, p.205:

> Then Colonel Buckley was called in to deliver his opinion. He was a Kerryman, considerably older than Bunworth. He sympathised with Bunworth on the loss of his men but did not agree with Bunworth's interpretation [that there was no way the attack could have been anticipated]. He said the soldiers, under orders, had been repairing a bridge which the local natives had destroyed. The local natives – Baluba of North Katanga – were loyal to the central government; the Tshombe government had tried to coerce them into accepting the jurisdiction of Katanga. A small-scale civil war had been going on in the area for months. Tshombe's forces had used the bridge to gain access to the refractory area and punish it by burning villages. The Baluba had accordingly destroyed the bridge. When they found the UN party engaged in repairing the bridge, they saw the UN as abetting their enemies and helping them to reinvade their home territory. So they attacked the party engaged in mending the bridge. 'It's very sad', concluded Colonel Buckley, 'but as a Kerryman, I can't see anything mysterious about it'. He was clearly referring to Kerry's experiences during the last stages of the Irish civil war of 1922–3.

Col. Buckley comments: I cannot understand that at all. I remember some things about the civil war in 1922 and 1923, but I was a child. I don't even know why he should say that. We must have talked about Niemba but I would have been the silent partner. I would have heard Bunworth talking about Niemba but I don't recall it. It was forty years ago. I cannot make sense out of that. I don't even know how Cruise O'Brien knew I was a Kerry man. 'It's very sad' – Jesus, it was terribly sad; we lost nine soldiers. That was the sad part of it. It had nothing to do with the civil war in Kerry, I can tell you. I know about things like Ballyseedy, and I know the place well. I remember during the civil war our house was off the road and republicans stayed there at night, time out of number. We were on the republican side. We were a safe house for them; it was off the road through a very narrow boreen. As a child I was aware that the war of independence was going on in Ireland. Given that, I was able to understand the circumstances that gave rise to the independence movement in the Congo. That is correct. I remember in our time, during the war of independence, that a foreign soldier in our country had no right to be in it. That would have been driven home to me pretty clearly. And I would have had the same view with regard to the Congo, which reminded me of it. I suppose it gave me a better understanding of what was happening in the Congo. That is so. In my time there, however, I did not see anything untoward happening. But I don't remember saying anything like that to Cruise O'Brien and I don't know where he would have got that. My father was not involved in the old IRA, but my uncle was.

There are five inaccuracies in Cruise O'Brien's book relating to me. First, is that we met in New York, when it was in Dublin. Second, that I was a much older officer than Bunworth. Actually, there were three years between us but we were cadets together and were promoted together. You would never know there were three years between us. I think he was born in 1918. Bunworth had nothing to do with the RAF, so that is an inaccuracy. The date of the Niemba ambush is also inaccurate. He gives it as January 1961, whereas it was in November 1960. He writes, in page 206: 'I found Colonel Buckley's interpretation wholly convincing.' Interpretation of what? If that refers to what happened at Niemba, I hope I didn't have an opinion. Bunworth and myself must have disagreed about something, but what about I cannot say. We had a different opinion on it. It certainly didn't break up any friendships. The discussions were in the Russell Hotel where we had lunch with Aiken. Cruise O'Brien says that there were four people present. I thought that the Minister for Defence, Kevin Boland, was there too, but I wouldn't be able to swear to it. General McKeown was not there because he had gone out to take over from General von Horn. That was about a fortnight before I left the Congo to come home. When I was passing through Léopoldville at the airport, General McKeown came out to meet me and say goodbye. I only did the one six-month tour of duty.

In fairness to Conor Cruise O'Brien, he had a most difficult job at that time in Elisabethville when Katanga was trying to secede from the Congo. Of the country's six provinces, Katanga was the wealthiest with all the minerals and real wealth. They wanted to break away and become independent. Dr Cruise O'Brien was certainly in the middle of that, but I was not there at the time. The 35th battalion was there and they lost men, including one officer.

I met a few Belgians out there in the Congo. I think they did an awful lot of good work there, but I would say they treated the native blacks – the same as everyone else did – as less than human, quite a lot. Any rows we dealt with would be women telling a black helper in the house that he was a monkey. In our time we would have to take the woman to apologise because they got very touchy about these things. I'd say they were treated very roughly. Nevertheless, there were many fine things done in the Congo. There were beautiful buildings, ranches and tea plantations. The Belgians had gone when we reached there. All that was left was an odd one here and there who seemed to be able to get on fine. They had no worries in the world about running their ranches. They were treated with respect because they treated the people who worked for them with respect. But I never saw any atrocities or anything like that. It was over and done with. There were a lot of beautiful houses in Goma lying empty. The furniture was left in them but they were never touched. We used to inspect them practically every day but nothing was ever touched in those beautiful houses. There was nobody in them but I don't know what the position is today.

I can only speak about my own term in the Congo. As far as I was concerned, we were justified in being in Kivu province. I think we did a marvellous job. We got the confidence of the people and there was peace and quiet all the time we were there. That was the name of the game; our purpose was to maintain peace and we succeeded in doing it. We never had a confrontation, although we were close enough but it didn't come to that. As everyone knows, Goma has been destroyed since with all the refugees and people being killed. So, has it gone back to a worse situation than it ever was before? I would say it has and that it hasn't recovered from that.

4

Reminiscences of a Colonial Administrator

From 1951 to 1960, Emmanuel de Beer de Laer worked as a colonial administrator in the north east of the Belgian Congo, near the border with Sudan. He looks back on his time there during the last decade of Belgian rule in Africa.

I started off in 1951 as an assistant administrator and was later appointed principal administrator of a territory the size of Belgium, called Mambasa. We were responsible for everything: justice, culture, even the roads. When something didn't work they went to the administrator. It was a forest region and there were only four roads that crossed in the middle; that was where we were based. We Belgians spoke French among ourselves but we spoke to the natives in their own languages. There were two main local languages – Swahili and Lingala. We didn't speak them very well, but well enough to get by. The population was almost entirely indigenous. There was no industry, nothing at all. The locals grew crops of manioc and peanuts to feed themselves and to sell to the colonialists. There were pygmies there too, but they lived in the forest and played no economic role in the area. The pygmies were illiterate and never went to school. I'm not sure if they were less intelligent than the local population but they had a lifestyle based on hunting, fishing and gathering edible vegetation. They were very primitive.

From 1951 to 1960, I worked in the eastern part of the Congo, not far from the border with Sudan. Six months before independence, there were round table discussions in Belgium with Lumumba, who'd been let out of prison, and the Congolese leaders. It was decided to grant independence to the Congo on 30 June 1960. At that time, those of us in the bush – far from any towns or cities – didn't know how much things had changed. We thought it was mad to grant independence so soon. And it really was mad because nothing had been prepared [for independence] and because we thought we'd still have another twenty years to bring the Congo to the stage where it could become self-governing.

Nevertheless, we had to prepare the way for elections to a new national parliament, but it was really farcical in our region because they didn't know what voting was all about. The chief in each village decided for them. That was the mentality. It wasn't a case of each African deciding individually. For example, Belgian judges came out to make sure that everything was going according to plan, and the Belgian administrators, like myself, couldn't go near the ballot boxes because it was thought that might influence things. When the pygmies came to vote they were accompanied into the polling booth by a Congolese native because the pygmies couldn't read or write and didn't know what was going on.

In our area the visiting Belgian judge was indignant and insisted that each pygmy went unaccompanied into the polling booth to vote. But the result was that the pygmies didn't vote at all because they didn't know how to go about it. That small example shows that everything was done too quickly. For six months we'd clearly explained the concept of independence, but they didn't know what it was. For our part it happened like that.

When I was leaving the Congo I had problems. They wanted to keep me there, they didn't want me to go. So, they didn't want to chase the Belgians away – at least, not those in the bush. After the elections things didn't go badly in our area. As I already said, the natives didn't understand what independence was. For example, there was a witch doctor in a local village who told everyone that independence would bring flood waters. He instructed everyone to build an aeroplane, which they did, in the form of a cross like a long crate so everyone could get in. And they remained inside the crate for three days after independence. We had to send the police to burn the crate to get them out, otherwise they would have died of sickness as they were also using the crate as a toilet.

The natives would believe anything, no matter what. They still hadn't grasped what independence really meant. And besides, they were afraid of the old chiefs who were much tougher with them than we were. Each territory had a certain number of chiefdoms – traditional entities governed by a chief with elders. These chiefs were often much stricter than us, so [when independence came] the locals didn't want to go back to the toughness they'd experienced under the chiefs before the Belgians came.

In our bush region the local population remained completely calm when independence came. It was a holiday but I had problems with local workers afterwards because they thought that after independence they'd suddenly get lots of money and live like the white people. The natives began to make demands of the white plantation owners who, in turn, came to see me. I negotiated a doubling of the natives' pay, and they accepted that until some of them rebelled because they wanted much more. Then the blacks took out their machetes but I held my ground. They didn't dare touch me because they had respect for the white man, but they said they'd come back that night. In the evening they drank arrack, a local alcoholic brew, and were therefore much more dangerous. Then I left, not over the border, which wasn't very far, but for a local administrative centre at Bunia. From there I was to establish communications with the town of Usumbura where other Belgians were grouped under the protection of Belgian Army paratroopers. In Bunia there was a company of black soldiers with white officers. Everything was going well until the rumour spread that the officers planned to destroy the camp with mortars; so the officers were taken prisoner by the soldiers. I transmitted this information to Usumbura and during the night the Belgian troops parachuted in to free the officers. Much of the action taken by the Congolese was due to ignorance or to rumours. The Africans tended to believe any kind of rumour that was doing the rounds.

The Belgian colonial authorities organised primary education for everyone. All the native children went to school for the first three years of primary, I think. They learned to read and write but the older ones couldn't do so. Such was the learning process and one could say that, by the time of independence, a quarter of the population could read and write.

At the start of colonisation [in the 1870s] in the Congo, there certainly were abuses which I won't speak about here. There were abuses during the [Second World] war too, because a lot of rubber had to be tapped from the forest for the Allied war effort. It was very difficult because if the trees were not tapped properly they could be felled by mistake. The natives had to produce a certain amount of rubber from each tree, and those who failed to do so were beaten with a bullwhip. That punishment was very hard but when I arrived there in 1951 it was no longer permitted, except for prisoners. They could receive three blows, which wasn't much but it was very hard all the same. That was when prisoners refused to obey.

With the benefit of hindsight, our big mistake was not to have trained them enough. We thought we had lots of time ahead of us and we didn't recognise that they were capable of being trained. For a long time there had been no university in the Congo and when one finally opened in the mid 1950s, many Belgians, including myself, thought it would become a breeding ground for nationalists. I didn't know it would happen so soon. Many Congolese became priests in order to study at university. Others were sent to Belgium, but not many.

Paternalism was the big defect of colonialism in the 1950–60 period. I believe we did a great deal for the Congo in many respects: there was no more famine, the people were becoming literate, and health care had been introduced. One is not ashamed of the work we did there. We made the natives happy despite themselves, but we never asked their opinion about anything. The white man decided everything, including what was good for the natives. It was the era of paternalism in all its splendour.

Half a century later, one can see that our action in the Congo was spoiled by two serious faults: first, a lack of confidence in the Africans and, thus, no training of an elite; and, second, paternalism.

Much earlier, perhaps twenty years earlier, we should have provided universities, encouraged the Africans to study at third level, and had a bit more confidence in them. As it was, Belgians held all the important positions, whether in the public or private sectors. The blacks only had subordinate functions. When independence came, suddenly, we had to give power to people who had never had it before. That caused many problems, as we should have trained many natives much sooner. Many could read and write but didn't go beyond that level. We should have allowed the Africans to attend secondary school and university much sooner.

Fundamentally, however, we had no intention of leaving Africa. In the mid 1950s, a Belgian priest called Fr van Bilsen drew up a plan that envisaged the Belgians would pull out of the Congo in thirty years time. He was criticised by everybody. Thirty years? It was crazy. Looking back, it would have been marvellous to have had thirty years during which to prepare for independence in the Congo. It was crazy to have only six months in which to do it. It was impossible to create the necessary institutions – based on European parliamentary models – as well as getting a rural population to understand all that in such a short timeframe.

The independence movement was born in the big centres of population, such as Léopoldville (now Kinshasa). Almost two years before independence, in December 1958, there was a revolt among the native population and quite a number of people were killed. The Force Publique – composed of black soldiers and white officers – had to put down the revolt savagely because of the killing and arson attacks. No one could have foreseen that the blacks would revolt because they wanted better conditions in the capital, Léopoldville. After 1958 there was little trouble elsewhere, apart from some unrest in Stanleyville, which was due to Lumumba. He was a minor postal clerk but had a strong personality. He had been imprisoned for robbing the till. But he was a good orator and had been influenced by Africans from other countries who were Marxists. One cannot say that Lumumba was a Marxist himself, but his revolutionary ideas came from abroad. And while we did not realise it, it's important to note that

5. Belgian colonial administrator, Mr Emmanuel de Beer de Laer, at a Congo river crossing, 1951

the whole of Africa was on the move at that time. Little by little, all the countries were becoming independent but we, the Belgians, thought we could keep the Congo for decades to come. It has to be said also that the Americans were very anti-colonialist and were leading a campaign against all the colonial powers. Therefore, international opinion was against us, but those of us who were working in the Congo couldn't see all that. We were working in a vacuum, unaware of outside influences.

The two independence leaders were Kasavubu and Lumumba. Kasavubu was head of the Bakongos, the ethnic majority in the Western Congo where he had enormous influence. Lumumba was very influential in the eastern province, of which Stanleyville was the capital. Under the influence of these leaders the mentality of the Africans evolved rapidly.

In the six months leading up to independence, the local population in my area was pro-European. But then a delegation of natives went to Stanleyville to meet Lumumba. He told them that if they joined his party they would no longer have to pay taxes and they would be free. They came back with boxes of party cards which they sold for 100 francs each. In the course of a single weekend the population of the territory became Lumumbists. That's how it happened, and during the elections they all voted for Lumumba.

As there were many illiterate natives, the ballot papers had pictures of animals representing the political parties. Lumumba chose a cow as his party's emblem but while the cow was highly thought of in the bush, it was regarded as an insult by the forest dwellers. I was administrator for a forest area where the locals thought the cow emblem was a white man's trick – so they refused to vote. We had to send a van 500 km to Stanleyville so they could ask Lumumba's officials if the cow really was the party emblem. They were told it was, so they travelled all the way back and voted for him.

After independence, Kasavubu was the president and Lumumba was the prime minister, but they did not get on. Lumumba was risking civil war. Many Belgians remained in the Congo as advisors and they backed Tshombe who was an army sergeant – the highest rank an African could attain. Tshombe hunted Lumumba and Kasavubu out of office and formed a government of young African graduates. Katanga, the richest province, seceded from the Congo under the direction of Tshombe who was supported to the hilt by the Belgians. Quite a few Belgian officers stayed in Katanga to support Tshombe's army. All the Europeans in Katanga – including my brother-in-law who worked there first as a security officer and later as a judge – were backing Tshombe. They were sure they could set up a state within a state. Then the Belgians living in Katanga pushed Tshombe to declare independence, thus sheltering the province from the anarchy unfolding in the Congo proper.

While the Congo was a complete mess at that time, Katanga had order, discipline and an economic boom. But what ruined it afterwards was that the United Nations, pushed by the Americans, did not accept the secession. The UN troops were sent in and, still now, I wonder by what right they did that. It's true that without Katanga the Congo would not have been viable because all the wealth was in that province, as well as in the Kasai diamond mines. But, all the same, it was interference by the UN in a country's internal affairs.

When the UN intervened I was no longer in Africa, I was in Belgium where public opinion was very anti-UN. The unanimous feeling was that the UN was getting involved because we were supporting Tshombe. I also felt that the UN should not get involved, that the Americans were behind it, and that Katanga should be allowed to exist separately. Since there had been a breakdown of law and order across the Congo, it was good to have an oasis of peace in Katanga. However, in hindsight, it is certain that the Congo would not have been viable without Katanga. At the time, Union Minière was involved in copper mining and other things there, and Katanga couldn't be allowed to separate from the Congo. I can say that now, but I didn't think so in 1961.

I was very happy in my job in the Congo because I had an enormous amount of responsibility, but I really don't regret having returned to Belgium. As a territorial administrator in the bush, I only saw my two elder children – then aged five and six – once a month because they were in boarding school. Later, when they went to secondary school, I only saw them once every three months. I recall that [after returning to Belgium] I didn't share my brother-in-law's view that Katanga should have become independent. I felt Katanga was a Belgian creation and that Tshombe was a creature of the Union Minière. That mining company produced cobalt, copper and uranium. In fact, it was uranium from the Katanga mines that was used in the first atomic bombs developed by the USA in 1945. And I asked myself, by what right were the Belgians continuing to colonise Katanga? You have to tell it like it was: Tshombe was a spineless type of person but his prime minister, Godefroid Munongo, was a really tough character. It was probably Munongo who had Lumumba killed, afterwards.

The UN decided that the white officers [in Tshombe's army] should leave Katanga. There was shooting and people were killed. Soon after Katanga broke away from the Congo, the UN troops were defeated by Tshombe's forces. However, UN reinforcements arrived and Tshombe's soldiers, under the command of Belgian officers, were beaten. Under UN pressure the Belgian officers had to leave Katanga and this led other Belgians to leave the province also. When the Congo regained Katanga, the Belgians were afraid and so they returned to Belgium.

I had left for the Congo in 1951 through idealism, to help the blacks. The missionaries also did civilising work. For a long time there were only missionary schools there, no state schools. A Belgian liberal party minister called Buisseret was the first to set up secular schools in Africa. Therefore, I think we really had a civilising influence but the colonialism, such as we administered, was unacceptable. It appeared to be acceptable at the time because we didn't have the same mentality as nowadays; it was an era of paternalism. With hindsight, one can see that colonialism was unacceptable. It was thought that everything European was good and so that culture was imposed on the blacks as if they had the same type of civilisation. But it was a very long time before we realised that the Africans had a civilisation, art and values of their own. We had not recognised all that. We wanted to impose our culture on them, and that was the fault of colonisation.

One cannot tell what would have happened if the colonialists had not come there. Before the Europeans came the Congo didn't exist as an entity. It was composed of a number of tribes and two or three kingdoms, like the kingdom of Msiri in Katanga. The Arabs took young natives from the eastern Congo into slavery. Unification and an end to slavery were brought about by the European colonisers. But a defect of colonialism was the lack of local industry. Apart from the breweries and cotton mills, all manufactured products were imported from Belgium. It was a long time before a small factory was built there for weaving cloth. We didn't want to industrialise the Congo because that would have meant competition for Belgian products.

The coffee plantation owners got subsidies to start their plantations but they had to wait ten years for the crop to mature, so the subsidy was insufficient. They had a tendency to pay as little as possible to the native workers. The businessmen also had a tendency to swindle the natives. What they did wasn't always very nice and, as an administrator, I often had to arbitrate in disputes between the natives and the colonial settlers. All that was the negative side of colonialism.

But administrators who had lived out there for twenty or thirty years were no longer capable of questioning what they had done in the Congo. When I returned to Belgium I did voluntary work for a non-governmental organisation called Frères des Hommes which did Third World development work. That helped me to question what we had done in the Congo and to see the value of these Africans – something I had not appreciated when I was there.

At the same time, one should not reject everything that we achieved in the Congo, notably: education, health care, the roads, commercialisation, the cultivation of coffee and tea, and the production of cotton. Many good things were achieved but – and this is what I call the mortal sin – we did it without the natives. We decided everything for them and it was a serious mistake not to have trained them at all.

There were three power blocs in the Congo before independence: the big companies; the Church, which was very powerful; and the administration. I'd almost say that the big companies were more powerful than the administration. For example, Union Minière in Katanga made immense profits and all the Belgians who had UM shares made big profits. It must be said also, however, that the UM workers were the best housed, paid and fed in the entire Congo. Belgium invested millions in the Congo, so one could say that it was normal to make a profit on these investments, but the share-out was unequal. The big profits went to Belgium. With hindsight, we should have taken less and devoted more for the benefit of the natives, but in 1960 it was too late. We did everything too late.

A long time before independence, in the 1950s, a delegation of twenty-five or thirty Belgian parliamentarians came out to visit our area. There were representatives of all the political parties: the Catholics had come to see the work of the missions; the socialists wanted to see how the workers were treated; and the liberals wanted to see how the private companies operated. They arrived in a convoy of official cars at Epulu, formerly a centre for training elephants. Some elephants remained there for the tourists. The politicians were put up in a hotel which had too few rooms so they had to share, which caused difficulties. Then the head of the Catholic group came to me and said: 'Tomorrow is Sunday so where is Mass being celebrated?' I told him that I went to Mass every three months and that the nearest mission was sixty km away. He insisted: 'We want to go to Mass.' It was terrible. But there must be a God for politicians because just then a native passed by and told us that there would be a Mass the next day, nearby at Mambasa, where I lived. So they would have their Mass. But there were prisoners at Mambasa and they worked on Sundays in their prison uniforms, carrying water for everybody. There was no running water there so they had to fetch it in big 2,000-litre drums which they carried. I didn't want the politicians to see the prisoners working on a Sunday, so I instructed the soldiers not to let them out of jail. The Mass was held in the open air so everyone could see the road. All of a sudden I saw the soldiers and the prisoners carrying the water drums on long poles, while others had pickaxes. 'What a catastrophe!', I thought. After the Mass the politicians wanted to know why the prisoners were working on a Sunday. Then I had a sudden inspiration and told them: 'The road you're due to take later today has collapsed. I've sent them to fix it so you'll be able to travel.'

I regret having gone to Africa – even if I was motivated by idealism – without having sought the opinion of the natives. I acted as I had been taught to, but I regret it now. My experience in the Congo would have been greatly enriched if I'd had the advantage of working with the Africans. We never asked their opinion, although when I went into the bush, in the evenings, I often spent three hours talking to the tribal chief about local customs, etc. So, one could not say that we did not know them but we never asked their advice about anything we did.

In the wake of independence reactions differed widely. Some Africans wanted the Belgians back because beforehand they'd had order and jobs. But it's true that they had also tasted freedom. It must be said that apartheid was very strong in the Belgian Congo; hotels were for whites only and the blacks weren't permitted to enter them. Six months before independence black people were allowed into hotels and we wondered what would happen. Nothing bad happened, however, except that the natives never wanted to drink beer in a glass, only from the bottle. But these are minor details and they were well behaved in the hotels. The bush natives didn't frequent the hotels, however, because they didn't feel at ease there. It was the more cultivated ones from the cities who did.

For the Belgians who returned to Belgium in 1960 or 1961 it was still relatively easy to find work. Former colonial administrators became civil servants after passing an entrance exam which was easy enough. It was tougher for the owners of tea and coffee plantations, some of whom tried to remain in the Congo. Many had to leave because of the troubles and they came back to Belgium with nothing; they'd lost everything, absolutely everything. A friend of mine went to Corsica. Some worked in industry or opened shops, while others went to South America or elsewhere but they didn't always succeed. It was tragic for them. Those who, like me, became civil servants, received subsidies for a while, so one couldn't complain. Nevertheless, the years spent in the Congo didn't count as years of service in Belgium for purposes of seniority or promotion, so we started over again at the lowest level. A colonial pension was paid, however.

Eventually, I got fed up listening to the old colonials saying everything was great in the Congo and nothing was right in Belgium. They were the same people who were always complaining in the Congo. For some years I tried to avoid them. I wanted to turn the page, to look forward instead of back. I was lucky to have come back to Belgium while I was still young. I got a second chance.

5

The UN Special Representative's View

In 1961, at a critical period for the newly independent Congo, Dr Conor Cruise O'Brien was appointed as UN special representative to the breakaway province of Katanga. In the following interview with Dr David O'Donoghue, he recalls his experiences as a key player in ending the secession of Katanga.

Q. You were the UN's special representative to Katanga.

A. Yes, I was the personal representative of Secretary-General Dag Hammarskjöld. That was my exact designation. I had been seconded from the service of my own country, the Irish Republic, at Hammarskjöld's request, to the secretariat. The idea was that I would serve there [in New York] for a little while and then he'd send me there. My own government didn't want to release me directly to the Congo, for a very legitimate reason: fear of getting too mixed up in the unpredictable and unpleasant politics of that place. So then, I went to New York, served a few weeks at headquarters, beginning to get an idea of what was going on, but only a fairly remote one.

Q. In talking about your role in Katanga, hindsight is a wonderful thing, but it's pretty clear that the whole Katanga secession was an attempt to create a Belgian puppet-state. Would you say so?

A. Oh, yes, absolutely. There was no doubt about that at all. The Belgian forces landed on, I think it was 9 July 1960, and Moise Tshombe was at that time provincial president of the province of Katanga. On 11 July, he went on television, with the commander of the Belgian forces which had just occupied Elisabethville, and in their presence he solemnly declared the independent state of Katanga. No outside power, not even the Belgians, ever recognised the independence [of Katanga] but it provided the framework on which they operated independently of the central government of the Congo.

Q. As regards your role, you went in on behalf of the UN Secretary General, the idea being to stop the secession.

A. Oh, certainly. In the earlier phase, before I came in, Patrice Lumumba was prime minister of the Congo. When he found he wasn't getting the Belgians out quickly enough, he decided to call the Russians in, which showed his innocence. The Russians sent him just enough help – mainly transport aircraft – to inflame the Americans against him. So, they, by various techniques, brought him down and then had him shipped off to Katanga where he was murdered. He passed through two airfields, Léopoldville airport and Elisabethville airport, which at that time were fully controlled by UN troops, who however were on strict orders from Hammarskjöld himself not to intervene between Mr Lumumba and his official pursuers. So they tortured him on the tarmac in both places while the UN forces, according to their instructions, looked on. I didn't of course know all that until later. I took it, which was the handout afterwards, that Hammarskjöld had had no idea of what was going on there, and that Andrew Cordier, his American deputy, took the law into his own hands, as it were. But that was all 'my eye'. We have Hammarskjöld's signature on the relevant orders, but I only found that out afterwards.

Q. Were you there at the time, in Elisabethville, when this was happening?

A. Oh no. I was brought in when a new [US] president had just taken over, J.F.K., and he had been more pro-African than Eisenhower. Eisenhower said that Africa should be left to the experts. The experts didn't include any actual Africans; they were the British, the French, the Belgians and the Portuguese. But J.F.K. was already on the way to shifting the balance and while that was under way the news came out that Lumumba had been murdered in Katanga. It was asserted that Godefroid Munongo, who was minister for the interior, had murdered him with his own hands. When he was taxed with this at a press conference, all he would reply was 'Prouvez le' [prove it]. It wasn't an easy thing to prove. This led to outrage among blacks and, in particular, among American blacks, and especially New York blacks. And they took over the galleries at the meeting of the Security Council convened to do this, and had a very effective and violent demonstration, so that they had to close the thing down. It was in that atmosphere that the Security Council passed a new resolution, which denied the legitimacy of the Elisabethville regime and authorised the use of force against it, if necessary, in the last resort. So it was under that mandate that I went in there and then a lot of things happened.

Q. Did you have a certain sympathy for Lumumba?

A. Yes, I did. I never actually met Lumumba but, though he was a man of little education, he had more education than most Congolese had. No Congolese at that time had higher education. The most any of them had was usually an unfinished secondary education. But he had quite a good secondary education and he was very bright, but naturally very ignorant of the political context in which he was operating. In a way, he was pushed in at the deep end. He wasn't, to begin with, any kind of extremist. Where things went wrong was at the official opening, by King Baudouin, the king of the Belgians, of the parliament of the supposedly independent government of the Congo. On the eve of that meeting the Belgian controller of the Force Publique giving instructions to his officers, wrote on the board, 'after independence – before independence', so it was clear who was still running the show. But then, against that background, the king came, and Lumumba came with a prepared text. He expected the king to be conciliatory and he would say how he would co-operate with the Belgian government, as head of the new Congolese government. But the king then made an extraordinarily inopportune speech in which, among other things, he praised 'the great benefactor of the Congo', this was Léopold II who, of course, was known not merely to the Congolese, but internationally, as a bloodthirsty tyrant.

Q. King Léopold II was Baudouin's great grand-uncle.
A. Yes, that's right. He had drunk in the family version of what Léopold II was like, which was ludicrously different from any reality. When he made this speech it was the decisive moment. Lumumba then threw away his prepared script and made an impromptu speech, violently anti-Léopold II and implicitly hostile to the whole period of Belgian rule. The rank and file troops were listening to this and when they heard it they realised that after independence was by no means the same as before independence, because no Congolese had ever before addressed any Belgian in such terms. They mutinied, drove out their officers, then got drunk, committed various atrocities and the whole of the Congo began to collapse into anarchy. It was then that the Belgians moved into the mineral-rich province of Katanga in order to safeguard their investments there. Following that, one had the events I have just described, leading eventually to the fall and death of Lumumba.

Q. Your autobiography, *Memoir: My Life and Themes* [Dublin: Poolbeg Press, 1998] makes harrowing reading, particularly where you quote Catherine Hoskyns's work, *The Congo Since Independence* [London: Oxford University Press, 1965]. Lumumba was beaten up on the airport tarmac [at Elisabethville on 17 January 1961] and forced to eat parts of his own speeches, with Swedish UN troops looking on. It almost has echoes of Srebrenica, doesn't it, where Dutch UN troops stood by when a massacre was taking place?
A. Yes.

Q. It's a terrible indictment of the UN, isn't it?
A. Well, the UN is never more than an instrument of the permanent members of the Security Council, by far the most powerful of which has always been the United States. Whenever the UN moves it's because the US has moved. We saw an example of that very recently, when the UN had been denying that it had documents which dealt with the handing over of some troops and/or some political people to control by their enemies. They had been in

denial of this in the last months of the previous administration in the United States, and now suddenly under this administration the UN have changed course and admitted that they did all the things that they had been denying they had done up to then. This comes as no surprise to anyone who, like myself, has been closely in touch with UN affairs. It's always the US that calls the shots and changes the shots when required.

Q. Are you saying that the US was behind Lumumba's assassination?
A. Not directly behind it, but they knew. The situation had been that the president of the Congo, Joseph Kasavubu, who became an instrument of the Americans, dismissed his prime minister [Lumumba] which he was not entitled to do under the constitution, but [Kasavubu felt] 'to hell with the constitution', by that time. It was at that point that Andrew Cordier the senior American went there. He told Lumumba that he would be safe as long as he remained in his house which was under UN guard, but if he left UN protection he would be on his own. He decided to chance it and make for Stanleyville where he had loyal supporters. When he was on his way there – he was travelling by car – he was intercepted by a plane. The Congolese army had no planes so it must have been laid on by the Americans with an American pilot. He was arrested on the way to Stanleyville and it was then that he was shipped off to Elisabethville where he was murdered. It was after the reaction to that that I came in.

Q. Do you think it's significant that in recent years the Belgian parliament set up an inquiry into the death of Lumumba?
A. It won't lead to anything.

Q. They must be concerned about it, though. It is haunting their past.
A. Yes, that's right.

Q. In your autobiography, you stated: 'Hammarskjöld had played a major role in seeing to it that Lumumba was handed over to the Katangans who promptly murdered him, as could have been predicted. But, once Lumumba was safely murdered, Hammarskjöld took a leading part in the international pressure designed to bring an end to the secessionist regime that had murdered Lumumba.'
A. Yes, that's right.

Q. That's rather convoluted and Machiavellian, isn't it? How do you explain it?
A. Well, he [Hammarskjöld] was convoluted and Machiavellian. That's quite a good description of him. He adapted to the political pressures that he was under, which were different political pressures as the situation in the Congo changed and the situation in the United States changed. He tried to fit each

successive state. For that reason, maybe, he began by backing Khiari and myself in preparing the overthrow of the Katanga government by force. Then when that began to go wrong militarily on the ground, he pretended that no such effort had ever been made and he issued a ridiculous document, which I deal with in my book *To Katanga and Back* (London: Hutchinson and Co., 1962) in the chapter entitled 'The Fire in the Garage'. He had an imperious use of language, so that words would mean whatever he chose that they should mean, rather like Humpty Dumpty.

Q. Did you get on with Hammarskjöld?
A. I got on quite well with him when we were in direct contact. He had picked me for the post, partly because he had read and liked a book of mine, my first book, *Maria Cross*, and partly on the horses for courses thing – he had followed my statements in the debates in the general assembly, which were more anti-colonial than most, or indeed any other, western diplomats' were. On the principle of horses for courses, when it was decided that he wanted to put pressure on Tshombe, he then thought that I would be the suitable person to go. Tshombe had said he would not accept any black or Asian representative, only whites. So I was about the only qualified white who met the requirement of what he actually wanted to happen.

Q. And from a neutral country.
A. Yes, that's it. He also knew me personally and even within neutral countries there are various things. I mean, my own immediate superior, Freddie Boland, and colleague, Eamon Kennedy, would have been much more sympathetic to Belgian policy on the Congo than I was.

Q. To try to put it in a nutshell, you were successful insofar as Katanga did not become independent and the Congo was maintained as an entity, but you can't have been happy with what happened afterwards, in the forty years since.
A. Oh, by no means. But the important thing in relation to the Congo was not what happened there but what the Americans wanted to happen. They picked Mobutu as a man whom they could rely on not to do anything that would be damaging to their interests, and above all to keep the Russians out, which was a high priority at the time. After that, they never took the idea of Congolese democracy or a Congolese parliament seriously. Indeed, it would have been rather hard to take them seriously. They opted essentially for a dictator that they could rely on. I mean, as they said about some other dictator in the area at the time, 'We know he's a bastard but he's one of our bastards' [laughs].

Q. It's terribly cynical, isn't it?
A. Very cynical, indeed.

Q. So, essentially that was why you resigned eventually. Did you feel you were dancing to the Americans' tune?

A. Well, I was being put in a false position. I was being made to pretend that things I knew had happened didn't happen. At one point in a state of depression and temporary demoralisation I came near to that by issuing a statement under pressure from headquarters to describe my mission in terms which I knew to be false. It was immediately after that that I recoiled. I said, 'I can't let myself be a prisoner of a lie for the rest of my life. I've got to get out of this, and I've got to go public on it. And there's no way I can do that except by asking my parent government, the Irish government, to recall me and then when they have recalled me to resign from their service.' I was then recalled. The recall was managed by the then UN secretary general, U Thant, who made a statement praising me for my stand on principle in Katanga – it was the exact stand I was being fired for [laughs] – so I could retire with the honours of war. They could not then say publicly what they had been saying privately – that I had made a mess of it. They put that in because they wanted to conciliate African countries, primarily Tunisia, whose support they needed.

Q. Essentially, what were the things you were being asked to say that you disagreed with?

A. That I had not declared the secession of Katanga at an end. I did declare that, but subsequently mended my hand. At the time I made that declaration, I had no idea that the UN was changing its ground. When I found it had changed its ground I briefly, and wrongly, played ball with the revised version. Then I came to see, 'You can't live with this. You can't stay, you've got to go.'

Q. But where did you make the declaration of the end of the secession?

A. Immediately after Operation Morthor – the dispersion of the Katanga government, the arrest of one of its members, and the flight of Tshombe to Rhodesia. It was from there that he declared the war was still continuing. It hadn't occurred to me – it ought to have – that Hammarskjöld would pull the rug from under me as a result of that.

Q. Did you declare the end of the secession verbally or in writing?

A. Verbally. I think it was during a radio interview.

Q. In Elisabethville?

A. Yes, that's right.

Q. But there was nothing in writing and Hammarskjöld then changed tack?

A. Yes.

Q. Niemba was one of the worst losses ever suffered by the Irish Army. What are your feelings about it? Do you have any doubts about the wisdom of sending white troops, including Swedes and Irish, into such a volatile situation immediately after the withdrawal of the white Belgian troops?

A. Well, he [Hammarskjöld] wanted to get some troops in in an attempt to restore the authority of the central government. The only troops he could get in at first were white troops. Later, of course, they were reinforced by Indian troops who became, in fact, the most efficient force there, and by Ethiopian troops. But that came later. In itself, the decision to send in European troops first, so as to get a toe in the door, was justifiable, but a lot of things that were done after that were not justifiable in my opinion.

Q. Including what?

A. Including being an accessory to the murder of Lumumba. That is the most heinous transaction of the whole lot.

Q. Within your remit in the UN, were you allowed to voice your feelings about that?

A. Well, you see, I didn't know. I had no idea. I accepted Hammarskjöld almost at his own self-proclaimed word. Certainly, I had no idea that he was such a Machiavellian person, as he actually was. I only found out that in stages as I went along, and not really fully until I had read Catherine Hoskyns's book which I didn't read until I had left UN service.

Q. That was 1965?

A. That's right.

Q. Of course, the Congo is worse now than ever, possibly.

A. Hardly that [laughs]. As bad as ever, I would say.

Q. If you could turn back the clock, what would you have done differently in the Congo?

A. There are certainly things that I would have done differently. After the arrest of the European officers and their expulsion from the Congo, which went off well, what I think now that we should have done was to put Tshombe's residence under UN protection with a UN garrison there. People whom Tshombe regarded as protecting him were people whose views he invariably reflected. His response to our expulsion of the Belgian officers, whom he had refused himself to expel, was 'Je m'incline devant l'ONU' (I bow to the United Nations). Well, if we had put a garrison there we'd have kept him bowing [laughs].

6

A Better Future Beckons Beyond Shadows
of the Past

BBC World Service journalist, J.J. Arthur Malu-Malu, was born in Kinshasa in July 1960, just two weeks after Congolese independence. Here he looks back at his country's difficult birth, the contentious role of foreign powers, and the Congo's unsteady development over the past four decades.

I would not say that everything about colonial rule was bad. The thing is that the locals couldn't enjoy their freedom properly. That was one of the bad things about colonialism. But to be honest during colonisation the country was equipped with some basic infrastructure and since independence in 1960 not much has been done to build more infrastructure.

I wouldn't say that I admire Patrice Lumumba. He was a great leader and a powerful speaker, but he didn't stay long in power. He became the prime minister of the Congo when it gained independence but he was sacked by the president two months later. So, it is very hard to say whether Lumumba would have been a good prime minister or not. People tend to speak well of him based on his speeches and his convictions. He had a certain sense of nationalism and a vision, an idea, of what the country would become and what exactly independence meant. Basically he can take credit for that. He was one of the fathers of Congolese independence, but I would not say he was the only one. To the best of my knowledge the first person who said 'We have to get independence' was not Lumumba, but Joseph Kasavubu who was the first president of an independent Congo. But Kasavubu was not a strong personality; he was keeping a low profile and was very modest. That's why he is not very well known, whereas Lumumba is known worldwide. The younger generation in the Congo do not know that the first president of the country was Kasavubu. He is not remembered as a great leader.

It is quite clear that Belgium, backed by America, wanted to get rid of Lumumba because he was labelled as a communist. In those days being a communist was something bad. They wanted to get rid of him because they thought that he would threaten their interests. Basically, a decision was made in Belgium [to get rid of Lumumba] but to achieve that goal Belgium needed some Congolese accomplices. That is why they recruited a number of people who were to execute Lumumba. It was not only Belgium, however, because you also had some of the Congolese leaders who were close to Lumumba in those days.

6. J.J. Arthur Malu-Malu, a
Congolese journalist working
for the BBC World Service's
African section

Moïse Tshombe was a leader who wanted the province of Katanga to become independent. He was very close to Belgium and was very close to the people in power in Belgium in those days. He belonged to a very rich family and of course the people in Belgium backed him. They wanted him to secede from the rest of the Congo because Katanga was one of the richest provinces. Belgium was behind that: they influenced him because they knew that would match their financial and economic interests. To a certain degree Katanga was a Belgian puppet state because there was a strong Belgian military presence there. The Belgians wanted to keep Katanga under their control because they knew that region was providing more than 60 per cent of the country's foreign exchange. Katanga was the main source of revenue for the country and that is why they didn't want to lose it. There were many Belgian nationals in Katanga who wanted to keep it under Belgian control.

I agree with the United Nations intervention in those days because the country did not have a strong, solid army. It was quite normal for those leaders to seek assistance abroad as no one else could intervene to get rid of the rebels. That is why they resorted to the UN. I find it quite normal. They helped the country to end the rebellion in Katanga. You could not achieve enduring peace with part of the country being held by rebels, so to me the UN intervention was a good thing. The rebels were behind Tshombe who also recruited foreign mercenaries who were paid by him and by Belgium.

After independence the country was torn apart and efforts were made to try to reunite it. People were going here and there around the world trying to bring those factions together. The conditions were right for a strong person, like Mobutu, to enter the picture and seize power because nobody was in control. The government in Kinshasa – called Léopoldville in those days – was so weak that it could not control many things. That made it conducive for the army, which was led by Mobutu, to grab power. Mobutu was a friend of the Belgians and was even said to have been working for the secret services. He was very close to Lumumba as well, which is a bit surprising because Lumumba was not a friend of Belgium. Mobutu was playing his own game: he knew that by befriending everybody he could advance his own hidden agenda. That was his strategy. He was close to the Belgians, close to Lumumba, close to everybody and knew how to play it.

My father was not a follower of Lumumba, he didn't like politics. He worked as a manager. In the late 1950s my father was among the people who backed a leader who is not well known outside the Congo, called Antoine Gizenga. He was a minister in the first government to be formed under Lumumba. My father was close to Gizenga who is about 80 years of age now and still lives in Kinshasa. Gizenga was one of those who decided to secede in protest against Lumumba's assassination. My father would not tell me much about Lumumba, but he told me about how sly and unreliable Mobutu was.

Nobody knows what would have happened if the United Nations forces had stayed longer in the Congo. But they left and the country could not achieve enduring peace. I do not want to talk about 'if's' because you never know what would have happened. But the fact is that the UN contingent pulled out of the country and the Congo could not achieve peace. Maybe if the UN had stayed longer an enduring peace would have been achieved and the country would have gone through a long period of stability. It is very hard to say what would have happened if the UN had stayed on longer. It is guess work about which I have no opinion.

Mobutu was able to hold the country together through strong leadership. To the best of my knowledge, we did not have ethnic strife such as we have today. But on the economic front, Mobutu embezzled a lot of money; we have to say things the way they are. He did nothing to improve the living conditions of his people. He was trying to help his cronies, the people who were around him. He made a lot of money but it was a very bad thing for the country. He was in power for thirty-two years but, to me, it ended in failure, economically speaking. At times, Mobutu acknowledged that he had not achieved his economic goals. Politically it was a failure also because Mobutu ruled single-handedly and the country didn't experience what they call democracy. We did not have free and fair elections under Mobutu, so the people could not speak their minds. In those days, if you tried to criticise the regime you would end up in jail. He was trying to silence everybody and make it impossible for people to speak their minds. That was sheer dictatorship.

Lumumba's style was different, he was not a greedy person. He didn't think that politics was a way of making money. Mobutu's view of politics was different – he thought it was a shortcut to fortune. If Lumumba had stayed longer in power things would have been different, but it could have gone either way. I wouldn't say it would have been much better, but Lumumba was not money oriented, whereas Mobutu wanted to pile up as much money as possible. He thought that in order to remain long in power he had to corrupt people and the money to be used for corruption had to be taken from the state coffers.

Mobutu was overthrown in 1997. He was kicked out of power after a rebellion led by Laurent Kabila. It took only six months for Kabila to get to power. Mobutu died a few months later in exile in Morocco. When Kabila took over people expected a lot of things. They were very confident and said, 'Well, at least Kabila is a Lumumba follower. We think he is going to correct some of the mistakes Mobutu made.' But to be honest, Kabila's reign was even worse than Mobutu's. He spent only four years in power but he exacerbated everything, not only the economic crisis he inherited from Mobutu, but added to that, the war as well.

The situation was very difficult, even for a strong leader, to try to improve things within the space of four years. But there was no indication that Kabila was willing, able or had the commitment to do that. There was no clear indication that he knew what he had to achieve or knew the kind of expectations the people had. He didn't do well as president, that's a fact. Under Mobutu, I would not say the press was free but people criticised him to a certain degree. When Kabila took over, however, there was no such thing as a free press. I could not tell you how many people were jailed during those four years. He banned political parties, outlawing them. He just decided those guys weren't needed. He had only one political party, which was his, backed by rebels who helped him to get to power. On the economic front nothing at all was achieved. The standard of living collapsed dramatically. He didn't do well as a president.

Laurent Kabila was assassinated in January 2001 and his son was hand-picked by a number of leaders – close associates of Kabila. Joseph Kabila is doing much better than his father. For instance, movement has been made towards peace. He has been able to give a new impetus to the peace process in the Congo and as a result all the foreign armies that have been fighting over there are pulling out of the country. He has held face-to-face meetings with the rebel leaders, whereas his father did not want to do that. On the economic front, Joseph Kabila still has a lot to do. But he seems committed and has achieved more than his late father.

Today the Congo is not divided along ethnic lines, as such, but people of Rwandan extraction are discriminated against. The nationality issue – who is Congolese and who is not – must be addressed for the country to achieve peace.

7

A Belgian Prosecutor Remembers

Jacques de Jaer worked as a lawyer in the Belgian Congo and remained on in Katanga during the turbulent post-independence period, acting as a state prosecutor in the breakaway province. He believes the UN should have backed Tshombe instead of toppling him.

At the time of Congolese independence (30 June 1960) the situation in Katanga was quiet but we knew we were in a difficult position because northern Katanga was mostly Baluba country. Most of the Balubas had good relations with Lumumba who was against the presence of Belgians and other white people in the Congo. There were other tribes in Southern Katanga, such as the Balundas, the Batshokwes and so on. They had good relations with Mr Tshombe who was the Conakat leader. He was also supported by the European people of southern Katanga, which was the richest part, containing very important copper mines, as well as cobalt and uranium deposits. Because there was a lot of trouble in the other parts of the Congo, we believed it was important to have – at least for a short time – an independent government to ensure a peaceful situation in the richest part of the Congo.

Later, Tshombe became prime minister of the Congo for a short time after April 1961, following the Tananarive agreement between Kasavubu and Tshombe. His intention was to make the whole of the Congo a peaceful country, not just Katanga.

The secession of Katanga did not last very long. It ended in mid-1962. Tshombe was indeed supported by the Belgian authorities whose idea was to try to have peace throughout the Congo and not only in Katanga. But we had to start with that part of the country because there was too much trouble in the other areas. When we tried to put Mr Tshombe as prime minister of the Congo our idea was to have a new and calm country – the whole Congo – as before. That was the real idea. Our idea was that, as soon as possible, the local authorities would take power but it was difficult to have black people who were able to govern. The great mistake of Belgium's policy was that they did not train local people earlier to govern the country. We had no local university people, no one who was able to take over in good conditions. It was a real mistake that Belgium did not train such people earlier. If they had begun to do so ten years earlier, in 1950, it would have been possible for local people to take over the government of the Congo.

7. President Moïse Tshombe shakes hands
with Belgian children in the mining town of
Kolwezi, January 1962. Just behind Tshombe
is Belgian lawyer, Mr Jacques de Jaer (centre,
back)

The United Nations was mostly anti-Belgian and also anti-Tshombe. The
UN wanted to support Lumumba but that was a mistake because Lumumba was
supported by most communist powers, including the eastern bloc. That was
probably why he was killed but by whom, I do not know. It is still a mystery.

In other parts of the Congo a lot of Europeans were massacred but not in the
southern part of Katanga. In June 1960, when I was in Albertville, I was frightened not so much for myself but for my family because just after independence,
there were rebellions in the Force Publique (regular Congolese army). At the
Congolo military camp there was an important rebellion so all the white people
there came back to Albertville. It was around 6th or 7th of July 1960 when the
women and children crossed Lake Tanganyika. They were afraid when the
black soldiers rebelled all across the Congo because they were armed and
dangerous. That is why most of the wives and children fled the Congo. It was
not too quiet for us either but we believed we had to stay. I believe that most of
the black people wanted us to stay because it was the only way to have peace.
They were afraid that if there were no longer white authorities there, a normal
situation could not be assured. We can see what the actual situation is now.

I understand why the black people may have hated the whites but it was not
directed against everybody; it was mostly against some whites who were usually not very friendly to the black people. They insulted them, calling them
macaques [monkeys] and that was the reason the blacks did not like them. But,
normally, the authorities, such as prosecutors like myself, tried to defend the
white people against other white people. That is why after independence I

stayed on working as a prosecutor and I had no problem. They accepted me in that role.

As regards the administrative authorities, the district commissioners or territorial administrators were black but always had the help of the former white officials, which they accepted. As an adjunct, the black officials always had white people acting as their advisors. This was accepted, at least in Katanga.

I do not accept that the Belgian Congo was a brutal society for black people, absolutely not. When I was there we were considered as negrophiles. Officials such as ourselves were only in the Congo for a few years, carrying out our functions but the white settlers, who thought they would stay there forever, were mostly against the black people. The estate owners were afraid of losing their properties. Officials like myself had no houses; we stayed in state-owned property and did not even have our own furniture. If we had to leave the Congo, therefore, we would not lose very much, apart from our jobs. We would have to find new jobs in Belgium. But the white settlers who planned to stay permanently in the Congo were against independence and against any authority being given to the blacks. Such settlers preferred the idea of Katanga seceding from the Congo.

Before independence, the chicotte [bullwhip] was used for discipline in the Force Publique but it was not a normal penalty handed down by the courts – it was never accepted as such. The death penalty existed but I never saw it carried out. I worked in the Congo from 1952 to 1962. Sometimes, I considered that my life was in danger but not to the extent that I felt I had to leave.

Other Congolese provinces wanted to be independent also, such as the Kasai region between Léopoldville and Katanga. The leader there was Kalonji. Not everyone liked Lumumba, therefore. Kasai was also a very rich province, with important diamond mines, and it wanted to be independent of Léopoldville.

The Niemba ambush: According to the information I had, which I vaguely remember, the Irish were patrolling the road from Albertville to Kabalo. That road was undulating enough and at one point they encountered a roadblock. They wanted to clear the barrier but in doing so they went too far from their vehicles in which, notably, they had also left their weapons. Just then, as I recall it, the Balubas arrived and surrounded the vehicle, so the soldiers found themselves unarmed and defenceless. Then they [the Irish soldiers] climbed an anthill area and were massacred on the spot – I don't know whether by firearms or side arms. Effectively, their mistake was to have left their vehicle, unarmed, to clear a roadblock.

Afterwards, I was in contact with the [Irish Army's] legal officer but, honestly, forty years on, I can't really remember any other details concerning the matter. I was in touch with them [the Irish Army] to put together the inquiry into what had occurred [at Niemba]. I had always had very good contacts with him [the legal officer]. I can't remember the details but let's say that basically everything always went along in a very friendly way with him. Things went very well between the Belgians and the Irish. But it was extremely difficult with other UN troops, for example the Ethiopians, Malians and others, because one felt they were anti-white, really against the whites. The Irish always wanted to be, and tried to be, much more neutral. Apparently, however, they were less neutral after the Niemba massacre. They were more collaborative against every possible attack by the Balubas or the ANC [Congolese national army] towards Albertville, that's for sure.

In March 1962, we were in Congolo and saw hoardings with the words 'vive l'ANC' [long live the ANC], 'vive notre prophete Lumumba' [long live our prophet Lumumba], 'vive Gizenga' – he was the deputy prime minister and from that region – 'vive Lundula' and 'à bas Kasavubu' [down with Kasavubu] – he was the president. So, the signs were not saying 'down with Tshombe', but 'down with Kasavubu'. We found these slogans on a school blackboard, written by the people who massacred the White Fathers who ran the school. The school was part of a big mission, a marvellous building. It was really unfortunate, all that.

In my opinion, Katanga did not succeed as an independent state mainly because of the UN troops. The UN wanted to occupy the whole of Katanga with the aim of reuniting it with the remainder of the Congo. When Tshombe became prime minister in Léopoldville, we said to ourselves, 'Yes, the Congo can be reunited, why not?' It was then that I asked my wife, Françoise, to come back with the children (they had left for Belgium earlier) because it seemed that the country was peaceful once more. We would try to restore it as a calmer and more unified region.

As regards the Katanga affair, the UN could very well have adopted a position which was less opposed to Tshombe and could have tried to avoid all that occurred subsequently, thus maintaining peace in the region. In my opinion, the UN was too anti-Tshombe, whereas Tshombe would certainly have been – and, for me, he was – completely valid as prime minister of the entire Congo. He would have taken care not just to defend Katanga particularly, but he would really have wished also to make the Congo a peaceful country – I'm certain of that.

Prior to independence, Tshombe had been an important merchant. He was a member of an important family, whose head, the Mwata Yamvo, was chief of the Balunda tribe. Tshombe was really someone; one had only to see him to realise that he was very affable and civilised with everybody, black or white. He had none of Lumumba's surliness. As well as being affable, Tshombe tried to make peace and reach agreement with everyone.

The last Belgian army officers left Albertville at the end of 1960. I, myself, left the city, on 10 January 1961, for Elisabethville. In April 1961, my wife came to Kolwezi with the children. She left for Belgium in December 1961. I was part of the public prosecutor's team in Kolwezi and stayed there until June 1962 when I left for Belgium. I would have stayed there but I had finished a three-year contract, which had commenced in June 1959. We always did three-year contracts and I had finished mine so I went back to Belgium.

8

Reporting the Congo under Fire

Dubliner Alan Bestic worked as a Fleet Street reporter in the Congo during the delicate post-independence period from 1960 to 1961. Here he looks back at some hair-raising incidents involving Irish troops with the UN peacekeeping mission.

In 1960, I had been working with *The People* newspaper in London as a freelance. I also worked for them in Dublin: my sole job was to stop them getting banned, which I did with great cunning, although they were subsequently banned. That was because Hannen Swaffer said that Sir Roger Casement was a homosexual who was a great patriot. The paper was being burnt ten minutes after it landed in Limerick. Casement might have been gay, but in those days Irish patriots didn't even go with women – they were our heroes.

I put it to *The People's* editors that I should go to the Congo and they said 'Yes'. I talked to a fellow reporter called Cahill who told me to go in the back door by Rwanda-Urundi. I flew out in an old Dakota to Albertville, which was the headquarters of the 33rd battalion. I think I had to go through Brussels; it was a commercial flight of Katanga Airways, run by Tshombe who was a puppet of the Belgian government.

The province had seceded and was then the state of Katanga with its own airline, but it was a mess. The Tshombe regime was totally corrupt, of course, it was appalling. Tshombe's troops were run by mercenaries – Belgians,

8. Fleet Street reporter, Alan Bestic from Dublin, chats to armed Baluba tribesmen in the Congo, 1960

Germans and God knows what – who were very dangerous indeed. They were swaggering around all over the place, pissed out of their heads, with large whores on their arms. If you angered them they would shoot you in a minute. It was an ugly scene. They were there for the money and to bolster up Tshombe and make sure that he won. I saw some mercenaries but I never spoke to them. My newspaper reports were about the Irish troops. Relations with the mercenaries were poor. You never knew what they would do; they could shoot you on the spot. They were all Europeans: Belgian, German and ex-French Foreign Legion guys. They would fight for anyone who paid them.

The Balubas on the other hand were violently anti-Tshombe. They sometimes mixed up the Irish UN troops with Tshombe's forces. It was very difficult to understand. Conor Cruise O'Brien's book, *Memoir: My Life and Themes*, tried to explain the Niemba business. There had not been a satisfactory explanation of it up to then. He said that the Irish troops were mending a bridge which the Balubas had destroyed. The Balubas assumed the Irish soldiers were Tshombe's mercenaries and that is why they attacked, which seems to be a reasonable enough explanation.

I arrived out there and was staying in a grotty little hotel in Albertville. I went out to the Irish Army's headquarters every day and went out on patrol with them a couple of times. My reports were published in *The People* in November and December 1960, and the *Daily Herald* in January 1961. I was not there at the time of the Niemba ambush, having arrived back in London some days beforehand. When I heard about it I immediately made arrangements to go back to the Congo again. I thought it was the beginning of a right bloodbath. The Irish troops had held their cool very well up to that, in the face of a lot of provocation. I knew these guys and I thought they might make reprisal raids. I was ready to go back but to my amazement the editor of *The People* said, 'No, it's not worth it.'

While in the Congo, I remember a wonderful character called the 'Badger' Burke of the Irish Army Medical Corps. He had fought for Franco in the Spanish civil war but apart from that he was a lovely fellow and a very good medical officer. He used to say, 'The 32nd battalion are said to be the cream of the Irish Army; I suppose that leaves us – the 33rd battalion – as the skimmed milk.' The 33rd battalion had all the trouble. Burke was scruffy, often half pissed and sometimes fully pissed, but a very good doctor. He told me that he had spoken to some of the witch doctors out there. He said, 'We could teach them a good deal but they could teach us a good deal, too,' which showed that he was a shrewd guy because their medicine was good and had been tested by time. The Badger Burke had the typical army humour and he could take a joke, too. One of the orderlies had put a little sign on the Badger's bedroom door which said, 'The Sett'.

I was out on patrol with an army captain. We were all sitting around talking and drinking tea when he said, 'Alan, did you hear about our brothel?' I said, 'No, I didn't.' He told me the story, as follows. There was a private – and you may not believe that an Irish boy would do this – but shortly after he arrived he commandeered a big warehouse. Then he commandeered a whole lot of army blankets, owned by the state, for this nefarious deed. He made a whole lot of little cubicles. He then went down and recruited a whole lot of young girls on the reasonable assumption that they were young enough to be free of disease. He then hired a very large Katangan gentleman and gave him a club, telling him: 'I want you to patrol around this warehouse, belting anybody above the rank of sergeant that you see approaching.' He put the girls in the warehouse and was in business. Later, I heard the brothel had been closed down after the padre castigated everybody involved from the pulpit and threatened to excommunicate them. But, looking back on it, the brothel story was only hearsay and may have been exaggerated a wee bit.

I met members of the Baluba tribe out there. The hairiest time was when I was accompanying a young corporal and a private. We had been out on patrol near Manono. We had two Indian traders with us who had gone out of town because it was being looted and bashed up by the Balubas. They left behind their servant, who was from Sierra Leone, in Baluba hands. They were going back to try to collect him. Captain Sloane thought it would be a fairly routine operation. We came round a corner in a jeep only to find our way back barred by a line of Balubas, and up in a field over our heads were tribesmen with bows and arrows. That was bad news because the poison of the mamba snake on the arrows would kill you in three minutes. You would have no chance at all – that is what killed them at Niemba. We stopped and got out. A young Baluba man there showed me a card with a facsimile of Lumumba's signature. He spoke well but drifted into realms of fantasy. He said they were going to take Elisabethville in six weeks and I said: 'What about Albertville?,' in which I had an interest as I was living there. He replied: 'We can take that any time we like.' As I said in one of my newspaper articles at the time, he wasn't talking such nonsense after all. I told him: 'But your bows and arrows will be facing submachineguns.' He replied: 'The bullets will bounce off us.' I asked: 'Why do you say that?' He said, 'The witchdoctor has blessed all our clubs, and so long as we hold those clubs we are impervious to bullets.' Here was an intelligent young man speaking far better French than I could – I was stumbling along – but he believed this implicitly. I think this is what happened: a lot of them were mission-school boys but in times of stress and trouble they went back to the old religion, which was probably better anyway.

It was getting pretty nasty when the young corporal said to me: 'Alan, ask the Balubas if we can turn the car.' He couldn't talk to them because he spoke no French. I asked them but they said we couldn't. Then the corporal said: 'Right, if it gets any worse, you lot get into the car and turn it. I'll hold them off with the gustav (his submachine gun) and jump on board as you go by.' He knew this would be impossible, of course, because he would be dead by then. I thought that corporal showed immense courage in a nasty situation and he never lost his cool. Eventually, they let us turn the car and we whizzed around the corner and hit a ditch that they had dug behind our backs. We got the Sierra Leone guy back, by the way. I've never seen a paler looking African than him. The Balubas were very unpredictable and you never knew how they were going to react and consequently we were pretty cautious. They were a most intelligent tribe. The Balubas were not mentioned in the trouble and strife that followed independence, including awful massacres.

Albertville was like an Irish country village. You'd walk down the main street and if you kept walking you were in the jungle. It wasn't a city, as we know it. Some of the Irish soldiers were learning Swahili, picking it up from the cooks who spoke it. The Irish guys in the Congo used to say, 'When we get back there'll be two schools in every mess – those who've been to the Congo and those who haven't, and we'll hate each other.'

I met some excellent Belgians working out there. They were ranchers and not only did they run their ranches very well but they went out to the villages and vaccinated them against disease. These guys were all under sentence of death because if the Balubas got them they would kill them; they hated the Belgians so much that it was said they skinned them alive. It was a very basic and brutal society, let's face it. Even in Casement's day, the Belgians were behaving pretty badly. The Congolese were well educated to primary school level; they could speak French, do sums and read and write, but no more. That was it. They were taught enough to take orders.

On one occasion, I was travelling in a train escorted by Irish UN troops. The railway was still owned by the Belgians. On board we had the general manager of the railway company and a couple of his aides. They travelled in what was called a first class compartment, but they had the blinds drawn because if they had been found they would have been killed on the spot, understandably in many ways. On the way back we were returning at about 5 p.m. on a little motorised rail car. Elephants would get out of our way, and Balubas, too, who were trying to stop us. They waved to us, they were probably just curious.

We stopped on the way back and the Irish soldiers gave us a crate of beer and some tea. We were jeering them, calling them country cousins and they called us city slickers. A few days later the poor bastards were dead. Niemba happened days after I got back to London, which is why I immediately made arrangements to return to the Congo. I never thought for a moment that the newspaper people would not want me to go back. I presume that I met some of the people who died at Niemba, but I don't remember any names. It was a very small group and they had all been out on patrol beforehand.

The Irish UN troops have never been given the credit they deserve for the work they did in the Congo. They were very good peace-keeping troops because they hadn't got a colonial background. They had sympathy for the indigenous people who had been under a pretty vicious heel.

9

The Scars of Niemba

Brigadier-General Patrick Diarmuid ('P.D.') Hogan[3] recalls a traumatic period of service in the Congo, with the 33rd Battalion, in the run up to and aftermath of the Niemba massacre.

In 1960, I went out to the Congo when I was O/C of the 20th FCA Battalion which included the regiment of Pearse. I spent six months there from July 1960 to January 1961. I went out with the 33rd Battalion – we were the second battalion that went out – Colonel Bunworth was the C/O of it, God rest his noble soul. We landed in Kamina, the big Belgian base in the middle of the Congo, which we were told at the time was a bolt-hole for NATO and European governments should parts of Western Europe be devastated by a Russian atomic attack, although I do not know whether that is true or not. It was a huge place with an enormous airfield and miles of runways, workshops, a hospital, billeting, and all sorts of family facilities as well. It had been built by the Belgians but probably with an input of NATO money. A short time after that we went to the town of Albertville on the shores of Lake Tanganyika – the eastern part of what had been the Belgian Congo. We distributed our companies and platoons around the province of Katanga which was our parish. It was a very rich province and the cause of a lot of trouble over the last forty years or so. It contained copper and the makings of atom bombs – uranium. It was very rich in minerals of all kinds. It was a veritable Pandora's box when they discovered what was in it. Since it was the richest part of the Congo – it also had diamond mines – it was the target for efforts by European commercial interests to hang on to power in it, so that even if political power were to slip away from them, they would hang on to economic power there. Of course, the new central government of Lumumba in Léopoldville – that is, the government of the whole Congo – tried to hang on to it. But Tshombe did a UDI [unilateral declaration of independence] on his own and separated from the Congo, but it didn't last very long. Within two years it was back in the Congo again as a result of United Nations action. But when we arrived it was still very much a Belgian province, if not in law at least in fact. The Belgian army was there in Kamina and we witnessed the evacuation of the last Belgian soldiers from the Congo. They came by train from Kamina to Albertville, as we had done. They took ship in Albertville to Usumbura, in what is now Burundi. From there they were flown home, although some may have stayed there until Rwanda and Burundi got their independence.

9. Comdt P. D. Hogan, third from left, flanked by Niemba massacre survivors Tom Kenny (left) and Joe Fitzpatrick (right)

We garrisoned that huge province. I hesitate to say how large it is compared to Ireland. I have a map which compares them, and Ireland would be a small spot in the middle of it. Our companies were very widespread – some of them we could only reach by aeroplane. They were an hour away by DC-3 or heli-copter. Some platoons were almost equally far away from our headquarters but we kept in close touch with them by land and air visits and by radio contact of course. There were huge distances involved. Our poor signal operators did their very best but, as with almost everything else that we had at that time, nothing was geared for the distances and the climate we had to endure. Our uniforms were a laugh – thick woollen suits. Our soldiers collapsed from the heat all over the place. Our signals equipment just was not geared for it at all. If you can imagine signals equipment bought for use in Ireland being put into a situation where you had to contact somewhere as distant as Moscow. I don't know if that is an exaggeration but if it is, it's not much of an exaggeration. That interfered with our operation but our signal operators were gallant boys and they operat-ed as best they could. They worked very hard at it indeed, morning, noon and night.

A few degrees from the Equator and we realised that we could not keep on with our soldiers rigged out in these ridiculous outfits. We were guarding a cot-ton clothing factory in Albertville so Colonel Bunworth sent our Quartermaster down to the factory and ordered so many hundred shorts and long trousers. We were able to get shirts from the UN. We were rigged out in proper shorts and longers, hundreds of which were made by this factory. We dished them out on a normal – not too generous – scale, as they were needed. We got very good Indian bush shirts from the UN, which were made in Hong Kong or Singapore and had been issued to UN troops for a long time. From then on it was a bit easier to move around.

We ran out of boots because the going was very tough on them. The UN sent us high boots, the tops of which were well up to your thighs. They had big thick rubber soles and inside they had two thick insoles. Attached to the eyelet at the top was a little booklet of instructions for soldiers on how to look after their Arctic boots! So we packed them back in the boxes they had come in to whatever cute bastard of a Quartermaster had sent them out to us from the UN supply base in Bari, Southern Italy. I suppose they had been left over from the Korean war.

Our troops got used to the life there very quickly. Just as it was a complete change from the equipment point of view, it was also from the food point of view. We were rationed out there and our Quartermasters were very busy, skilful and devoted. None of us went hungry during the six months there; we had local supplies. We got a ration pack that was made in Dublin on the design of two officers in plans and operations at Army Headquarters. It was a gem. The officers went to Dublin firms and asked them to make up stuff, which they did at no cost. The government would not give them any money for doing this because they were not supplying anything on the basis of three tenders carefully annotated and submitted on blue paper. The Gateaux company – or it might have been Johnston, Mooney and O'Brien – made up a fruitcake. They gave the army a complete night's run of their bakery, for which they got sweet F.A. Dublin firms rallied round even when there was no notion of anybody going to the Congo or anywhere else. This was purely advance planning which was very well done by our Army Headquarters, to whom I seldom give any encomiums whatsoever. The packing of blankets and tents was first-class. It was all pre-planned and ready to go into operation when the bell sounded. The ration pack could be sold or exchanged in the Congo for other foreign items such as tyres, petrol and batteries – anything you happened to be short of at the time. It was better than hard currency, not that we did it on a wholesale basis. But occasionally it swung things in our favour when we were scrounging for something from another country's unit.

The battalions that serve abroad today are superbly equipped because of influences that began in the Congo in 1960 and which sparked off this improvement in equipment. The Government and the Army saw – although the Government saw it less quickly – that you couldn't send soldiers abroad like this any more. They had to have good arms for one thing. We went abroad with Lee Enfield rifles. Service abroad brought about an improvement in weaponry, clothing, tents, food, transport and radio equipment. When units go abroad now they have as near as possible to everything, but that process began way back then. We did not do it all – the successive missions improved year after year – but the Congo was the catalyst in more than material things. It was a catalyst for the spirit and morale of the soldier and of junior and senior NCOs, whose *esprit de corps* got a tremendous shot in the arm from this business. Their sense of responsibility went ahead by leaps and bounds. Corporals at home might have been in charge of a barrack's guard for 24 hours. They were very closely supervised with the orderly officer visiting three or four times during the night and maybe five times during the day, and the sentry outside the guardroom constantly under the eyes of the commanding officer driving in and out to his lunch. But in the Congo all that was gone. A corporal who was an hour away from his company headquarters, would be in charge of eight or ten men. He had to be responsible for their discipline, food and the running of his little post. Suddenly he grew into the job most magnificently. The same was true of the senior NCOs and junior officers who now found themselves alone facing responsibility and arriving at decisions they had never had to face before. In that six months of experience in the Congo, the Army suddenly grew up to an amazing degree, and thereby established a pattern of training from which future battalions also benefited. And like the clothing and the radio equipment, this was a geometric progression from one year to the next.

I was a Commandant, second in command to Col. Dick Bunworth, and most of my time was spent at battalion headquarters in Albertville. Bunworth's command of the battalion was extraordinary; he was an unusual man who was held in tremendous respect in the Army. I know of only two or three men in the Army that nobody ever says a bad word about, and I am not one of them, but Bunworth was. I am not standing back from this problem, I speak as a very close friend of his so you can take it that I admired him. He was a man whom I held in great respect and, indeed, affection. He commanded the battalion in this way which was quite a feat. Nobody in the battalion, down to the youngest private, wanted to do anything that would hurt Bunworth or displease him. One got the feeling that apart from being there to serve their country and the United Nations mission, they also felt a deep sense of personal duty to him. They felt that maybe their number one job out there was to act in a way that would please him and would not cause him distress in any way. This is my impression and it is one that was shared by many officers whom I spoke to on the subject.

Our disciplinary record out there was marginal, almost nothing. We had very few charges. We had a few fellows who perhaps were not suited to go out at all – suddenly the heat and loneliness of being away from their natural habitat, friends and family, got them down. One fellow wanted to express himself one night looking out through a window, so he ran up and down the billet and put his fist through all the windows. In that way he felt better. I don't know why, but there you are. He didn't injure himself doing that, at all.

We were billeted in a convent of the White Sisters who had a school there. Dick Bunworth said the Reverend Mother had only one interest – educating young black children. If she could do that better by getting us out, that suited her book all right. She wanted one classroom after another back, which we were using as billets for soldiers or officers. She was at us for six months, telling us

to give back rooms. The school was still running, but in straightened circumstances. We mentioned this one day to the assistant civilian administrator who
was a Belgian – the fellow in charge was a black Congolese – and he said in his
marked Flemish accent: 'Ach yes, it is very difficult for soldiers to cohabit with
nuns!' The Reverend Mother was Belgian; they were White Sisters, in their rig,
name and race. The school children were natives. There were no white boys or
girls there. I suppose the white children went to school in Léopoldville or
maybe at home in Belgium.

I never heard of any atrocities by the Belgians. One police officer there had
a fairly quick trigger finger and he shot a few people up on the frontier with
Kivu. But it could be said that the fellows from Kivu were invading Katanga
across the river which was the border and he was a policeman sworn to defend
Katanga so it wouldn't be too blameworthy. I never heard of anything they did
out of the way. When the troubles started, the local Congolese did some terrible things to Belgians that they captured – dreadful, unspeakable things. That
was after we arrived. Before we arrived, in July 1960, the Belgian Congo was
a very well governed and peaceful institution, as colonies go. I am not going
into the morals of having colonies or anything like that, but it was very pro-
gressive. They had a wonderful primary education system that was able to produce stationmasters and postmasters. All the lower levels of administration,
which in many parts of British Africa were filled by white people, were filled
by local Congolese. The long trains bringing raw copper to Rhodesia were
pulled by huge diesel engines, which even then must have cost millions of
pounds each. They were driven by black drivers and when they got to the border the black driver would climb down and the North Rhodesian driver, a white
man, would climb up into the cab. To my mind, that distinctly delineated the
difference of approach to the natives in both administrations. The Belgians
believed they should educate everybody well at primary level, then educate
people who could be educated at secondary level, and that the university level
would come out of that. When we arrived the university in Léopoldville had
just been opened and had taken in its first students. They believed that this pyramid of talent would grow naturally. They did not believe, as the British did, in
taking the chief's son or nephew out and sending him to Oxford or some place
like that. The British way produced some results quickly but the Belgian
approach was one of *festina lente* or hasten slowly. But they did not get time;
the winds of change – as old Macmillan said – were blowing through Africa.
They blew the concept of a pyramid of education away in the Belgian Congo.
You could find blacks in positions of moderate responsibility, at a middle level
such as postmasters and station-masters, which many people of advanced education in our own country are glad to occupy, and why not?

Why did it all go wrong? An awful lot of the blame can be placed on one
man, Lumumba – an unbalanced leader who got into a position of power. Also,
they did not prepare. I do not know how their local government worked; I
wouldn't say there was any. I just don't know. But you would have had some

form of legislative preparation within a lot of British colonies, whereas you did not have that in the Congo. I think that was the case, although I am not speaking from knowledge of the subject. Also, they had a very small white military presence, which was probably too small. The Force Publique was a very large institution but I don't think there were any white soldiers in Albertville, which was a big town by African standards, on the shores of the lake facing towards English Tanganyika and Usumbura. Maybe that had something to do with it, although I don't know.

There was a transfer of power from Belgium to a local government in June. It was at the transfer ceremonies that Lumumba made the terrible speech in the presence of poor King Baudouin which sparked off the whole rising. There was an election and government was handed over to the blacks. The dogs in the street knew things would not be as they were before. A couple of years ago, nobody in Pretoria said that things would be the same as they were [during the apartheid years].

As regards attacks against the Belgians, there was one instance of it in the north of Katanga, while we were there; a fellow was captured and I will not go into it but I would not like to have been in his position. Such activity was not widespread, however. You couldn't protect everybody in Katanga. We protected economic targets which would allow the economic life of the country to go on. We protected electricity stations, factories and the collection of cotton so that people would be able to live normal lives. In some places where our troops were, that very fact protected lives physically.

As to what went wrong in Niemba, the Balubas there were led by an ex-ANC sergeant major who was looking for arms and ammunition. He knew what he was doing. Any talk about our fellows being confused with Belgian mercenaries, I think is a lot of balderdash. This fellow was out to get arms and ammunition and he got them. A Baluba youth was wounded by an Irish soldier sometime before Niemba and it might have caused unrest but he was looked after. One of our men shot him by accident in the village of Niemba, but he was looked after. His wound was very slight and did not put him into hospital. It was a head wound. He was from the Niemba area. The Balubas mostly came down from the north where they were more numerous than they were around the battalion headquarters, or even Niemba which was far to the west of battalion headquarters. I would not see any connection between the attack on our troops and the slight wounding of this young fellow. There were no other incidents that might have provoked the attack at Niemba. There was nothing else that one could remotely associate with it, as far as my memory goes.

Our company up in Manono, where the Baluba rebellion began, were guarding places in the town and around it. I am sure the Balubas didn't like that. They had the effrontery to mount a guard on our company headquarters office up there, but we put the run on them for that. We went to the leadership and said that this was not acceptable, and they withdrew them. But they got on all right with our company up there in Manono; there were no fights, no hard words or anything like that, to the best of my recollection. Our company steered a very careful course there, avoided any incidents or possibility of incidents. I could fairly say that they didn't exacerbate the situation in any way. So, again, it would not be reasonable to associate that [with the attack at Niemba]. In fact, it was the other way around; they were so friendly with our troops up there that one would say it was a bad show on their part to attack their comrades down in Niemba, and kill some of them.

There was one servant in the platoon headquarters in Niemba who wanted to leave, and they paid him and let him go. That was just days before the Niemba ambush, and we felt afterwards that he had been told that trouble was on the way. Hindsight is wonderful, but we didn't associate it with any danger. We were not aware of this at battalion headquarters, but the people in Niemba – Lt Gleeson and his Sergeant – didn't attach any particular significance of foreboding to it. It's easy to see afterwards that he was told to get out, of course.

As regards my own recollection of the events leading up to Niemba, we went out two days before. I had a strong patrol with me from headquarters – and I stayed the night with them in the platoon in Niemba – to take off with some of their fellows and all my fellows down west to try to make contact with a similar patrol coming from Manono, eastwards, to meet us. We rose very early the next morning, in darkness, and took off. We were constantly moving roadblocks on the way to the river. That was not unusual; roadblocks had been building up around Niemba for a week or two before that. During the first week of November [1960], Lt Gleeson's signals contained a note of concern about the number of roadblocks around Niemba. This was an effort to see if one could clear them completely – we did not know these were there between us and the river – to see if we could go so far as to make a junction with the Manono patrol. But we had to clear a good number of trees which had been knocked down in our path. We were on motors and trucks. We eventually got down to the river Luweyeye and the bridge there was destroyed. We did our best to get a box of timber across but it was beyond our resources. We had a small engineer party but they didn't have very much equipment with them. They had buzz-saws but nothing that could put a box of timber or trees across the river. So, we had to turn back. While we were there we spotted natives in the bush. One of the lads went in and fetched one of them out. He was very small. Somebody said they saw pygmies in the bush. The fellow that came out wasn't a pygmy as far as I recall. We gave him some of our rations. The general impression was that they were the party that hit our fellows the next day, after we had fed them. They did not display any anger towards us that day. Our interpreter said they were hungry so we gave them food. As far as I remember, they didn't have weapons with them, although I am not sure. I can't recall that.

We returned to the village of Niemba that evening. I think I went home to Albertville the same evening with my patrol. We sent a message to the lads to go down the next morning to see if they could even get as far as the river. Well, they did and they were hit, and that was that. The patrol that was ambushed did not have an interpreter. That night back in headquarters, we got a message from the sergeant that Lt Gleeson had been left behind, that they hadn't come back. We gathered a big patrol together and sent them out. It was a very long journey out there – it was over three hours, and sometimes it would be five hours. They travelled all during the night and got out there. The next morning I was awakened by the orderly officer who said: 'We're in trouble.' The officer in charge of the patrol had gone through Niemba and went out as far as the bridge. He found spent cartridge cases but the lorries were gone – they had stolen them. He didn't find bodies; there was nobody there. Col. Bunworth immediately ordered me to collect a big patrol and get out, so we did. We got there in the late afternoon. Niemba village is up on a hill and the river winds its way around it. It is a big river, the beginning of the Congo river which flows out of Lake Tanganyika. It is not the river where the ambush took place. We found the bridge. It was not destroyed, it was too big to destroy. The river there would be bigger than the Liffey at Islandbridge, much wider. It was an iron bridge with wooden struts on it. They had removed a lot of the wooden struts and had thrown some of them into the river. We retrieved them and advanced. We heard a shot from the village or near the village. We were in radio contact with the officer who had gone out the night before and he told us that there was a big mob of Balubas in the railway station, which was at the other side of the village down at the bottom of the hill. We crossed the bridge on foot and spread out on each side of the road up to the village. We advanced through the bush upwards and didn't meet anybody. The lorries were on the road. When we got there I sent an officer, Dan Crowley, towards the bridge and had an anxious time waiting for him to come back with his patrol. It was dark and he had to move blocks going out and coming back. He brought back some bodies, I forget how many. We then decided to go out again the next morning. I got a radio message from headquarters to go out the next morning to collect the rest of the bodies. We settled down for the night and organised the defence of the village. We placed lorries facing out the roads from the centre of the village on the approaches. We had sentries up on the roof of the only decent building in the village. Late that night they shouted that somebody was approaching in the bush. Someone shouted a challenge but it wasn't replied to and firing broke out all along the line. Then the lights went out. A round had broken the wire from the electric generator, so I ordered the lights on the lorries to be turned on. I ordered people to stop firing. I and the other officers dashed out from the little house. After a few minutes, it transpired that one soldier had been killed accidentally by one of our own men. He was walking on the veranda of the house and there was a soldier sleeping inside who was wakened by the shooting. He woke up and saw this figure with a weapon going by the window. He grabbed his weapon, fired and shot him in the stomach. We had a doctor with us who did his best for him, but he was very seriously wounded in the stomach. The doctor said that if we did not get him in to Albertville he would die in Niemba if he was left there much longer. We organised a patrol and got a small lorry and put mattresses on the floor of it. We put him in there, covered him up and put a medical orderly with him to take his temperature from time to time. The doctor went with him and a fairly strong patrol to see that he got through. When they got down to the bridge the planks had again been thrown off it, so the officer in charge put them back. I radioed Albertville to send an ambulance and a party out. They met but the poor boy was dead by then, or very shortly afterwards, I'm not quite sure which.

The next day we got up very early and started off. We had members of the Ethiopian battalion with us, from Nyunzu. They stayed with us during the night and came with us the next morning. They were very useful indeed. We were a strong patrol – about ninety or so when we started off down to the bridge. I forget if we had roadblocks to remove or not. But when we neared the bridge, Captain Crowley, who was in the lead lorry, stopped the lorry and the whole convoy stopped. He got out. I saw a soldier on the side of the road, so I jumped out of my lorry and went forward to join Capt. Crowley who had not spoken to the chap; he had hardly reached him. I said: 'What's your name?', and 'I'm glad to see you.' He said: 'I'm 57 Kenny, sir, and I'm glad to see you.' It was a good one-liner. We took him in the lorry and drove back to an open place in the bush where a helicopter could land. I radioed to get a doctor out. He was badly wounded. The back of his head was black with blood and he had two arrows hanging out of his behind. This was over forty hours after the attack. He had been rambling in the bush. He was very low but he had plenty of old guts in him still. He wasn't beaten by any means. Before the helicopter came, our medical orderly and the Ethiopian medical orderly removed the arrows from his buttocks, which was quite an operation. It must have been very painful. We gave him a cigarette, that was the only medical assistance we could give him. He withstood it very well indeed. He is a fine fellow, Kenny. The helicopter landed and we had him sent back to Albertville.

We continued on and came to the bridge. We instituted a search and got the remainder of the bodies during a long, hot, terribly tiring and frustrating day. The Ethiopians were good trackers, better than we were, and were a great help in finding the bodies in the bush. I remember one of them standing up, sniffing and pointing to a spot. And that's where the body was.

I omitted to say that Private Joseph Fitzpatrick was found the night before and taken in with the patrol that came back after dark. They were the only people who were saved from it.

We got all the bodies except one. When I found Kenny he said: 'Trooper Browne saved me, sir. He [Browne] said: 'I have a gustav and I have some ammunition. You are married and have children. Go on, run and I'll hold them off.' He did and when I went away I could hear them coming upon him. He fired at them but I heard them coming upon him and knocking him down, beating him and killing him. *So, we presumed at that stage that Browne was dead. We couldn't find Browne's body. We stayed and searched but we couldn't find it. We went back to the village where Col. Harry Byrne had come out. He was the brigade commander in Elisabethville, the provincial capital. He had come to Albertville when he heard of the business. He came to the village with some other officers from brigade headquarters. He ordered us to evacuate the village and go home, so we did. The bodies were packed off in a helicopter and we hauled down the flag and moved. We packed all our stores and took off. It was a sad business. Sadder was the fact that Browne was alive – a fact which hasn't added to the gaiety of my life since, I can tell you. He travelled some miles through the bush, as Kenny did, but he didn't come out onto the road as Kenny did. Kenny was lucky to find the road. There was no road right or left of it for twenty miles. He was bloody lucky. But poor Browne didn't find the road. If he did we would have picked him up. He went back towards Niemba – whether that was by design or by accident – through the bush. Some days later on he was lying outside a village and some young women came out collecting firewood. By signs, he offered them money for food. They went back to the village but instead of bringing out food they brought out young men who beat him to death. We found his skeleton two or three years later on. I have a photograph of it, indeed. His remains were brought home and buried with full military honours in Dublin.

From Albertville, we mounted an operation to the hospital in Manono where B-company was. We raided the hospital and took out of it all the [Baluba] fellows who were wounded. Apparently the [Irish] lads felled twenty-six of them at the first volley but they never got a second volley off. It was because they never had to shoot before and they were too close to them by the time they fired. It goes with the job that you don't fire unless you are quite certain that you have to in self-defence. They had to wait until the first arrow was fired and then they fired but they were in amongst them in no time at that stage. I found that some of the blacks were buried on the site. I have a photograph of a heel sticking up out of the ground. So, they buried some of them but the wounded they took away with them, of course. We sent an aeroplane to Manono and had lorries from the company waiting there. We did a dash to the hospital, snatched them out of the beds, put them into the aeroplane and brought them back. There is a photograph of those fellows being taken out of the aeroplane in Albertville where they were handed over to the civil authorities who tried them. That is where the stories of the ambush came from really.

The fellows we took out of the hospital were tried in 1961. There must have been no death penalty in the Belgian Congo. They were sentenced to imprisonment; moderate sentences of three, four or five years. They were not punitive terms of imprisonment considering the deaths of our nine men. I heard that twenty-six Balubas were killed. That figure must have come from the fellows who were tried. The patrol had some Bren guns, of course, and the NCOs had gustavs, but who fired what we'll never know.

As to whether Niemba could have been avoided with better intelligence, reading the signs of the unrest among the Baluba in the days before, yes – we could have stayed at home in Dublin. An about turn at the airport would have achieved that, but what were we doing there? We had a mission to keep the means of communication open and to protect the civilian population. To achieve that we guarded factories and things like that. Of course it could have been avoided by doing nothing.

We had been carrying out patrols for four months. It was not the first time a patrol had gone out without an interpreter. We had been sending patrols out under people, mostly under officers, into the wilds. They met parties of Balubas and they'd put up their hand and shout 'Jambo [peace]'. The Balubas would come forward and chat if they [Irish troops] had an interpreter, or if not, they'd exchange food. The massacre was a shock. I will live with the fact that we failed to find poor Browne, who was alive. That's a bitter pill, very bitter. For the rest, we were there as soldiers to do things. We were not interior decorators.

10. Irish UN Troops search the railway line near Niemba after Balubas derail a train

---

* This version of events – wrongly attributed to Thomas Kenny by P. D. Hogan – is naturally disputed by Mr Kenny. In any case, the remarks are contradicted by Hogan's admission that Trooper Anthony Browne died 'some miles' away in a seperate incident. The Irish Army's official report on the recovery of Browne's remains in 1962 (see Appendix V and VI, pp. 214–7) supports Kenny's version of events.

10

Reliving the Nightmare

Thomas Kenny still lives with the nightmare of the Niemba ambush, which, he claims, could and should have been avoided by adopting a more cautionary approach. He says the ominous warning signs were there to be seen prior to the massacre.

A couple of months before the Niemba ambush, a Baluba chief's son suffered a head wound when challenged and shot by one of our men on guard duty. A 24-hour guard was placed on the Baluba youth while he was in hospital. When he left the hospital, word was sent to the Balubas. To my mind there was going to be retaliation – because the Belgians had committed atrocities and the Balubas felt the Irish troops were out to do the same – and Commandant P. D. Hogan was told all this by the Army or UN administration.

To my mind, the ambush [at Niemba on 8 November 1960] did not happen, it was caused. Comdt Hogan was told to withdraw the troops from the area but he did not. We were considered by the Balubas to be aggressors, like the Belgians. The Balubas were trying to get rid of one lot, and then here was another lot. There were, reportedly, signs of trouble. On the day of the ambush, a Baluba cook who worked in the camp kitchen asked Private Gerry Killeen if he could go early.

11. Tom Kenny holds one of the arrows that wounded him at Niemba

I had been in Niemba about a month before the massacre, when a train was derailed there. The airports were closed at the time. Half a mile of track had been taken up. Of the approximately 100 Belgians on the train only five were found alive – one nun and four Belgian women. They had been gang raped. I was asked to get them blankets and a bag of fruit. The situation was bad. In fact, I had forty-six blankets but only needed five. The women's faces were covered with blood and the nun's clothes were all torn.

On another occasion, I was on a train and we were surrounded by Balubas. As we slid open the door, they were looking down the barrels of our Bren guns. The [Baluba] chief was standing there naked with a rope around his waist and leaves on his head – like Caesar – holding a stick. They took the water from the boilers so the train could only go at five miles per hour.

At one stage we were billeted in a house in Niemba. The Belgian or South African owners had been massacred. Behind the wardrobe was a wall covered in blood. And the mattress had been turned because it was covered in blood. They were a couple and two kids who had been killed and buried in the front garden.

The day before the Niemba ambush, Comdt Hogan had been on a patrol with fifty soldiers, including Fitzpatrick and Gaynor, and the bridge was intact. It was a vast region. The working party had gone out to keep the roads open. Gerry Killeen asked me to carry out a tray of sandwiches to the patrol when they came back.

Eleven of us went out on patrol the day of the ambush. One of the men was an Army chef who was bored dishing out meals. We were supposed to meet up with another patrol because they had come across too many obstructions on the road. The biggest obstruction on our side of the river was the [damaged] bridge, but there were hundreds of trees cut down on the other side. My first inkling that something was wrong came three miles before we got to the bridge. There was a headhunter's fur hat nailed to a tree. The first words I said to Lt Gleeson were: 'Did you see the headhunter's hat nailed to the tree?' But he dismissed the idea as if I had never asked the question. Only Fitzpatrick heard it because he was sitting in the same car. They told me that I was a jinx, but I was anticipating something. It was not as if someone forgot the hat – it was deliberately nailed to a tree.

Later on when we got to the bridge, the Balubas were there in strength. Nobody knew they were going to fire. They were shouting and screaming. We could only shoot in defence of our comrades' lives, or your own life, in self defence. You had to wait until something happened to return fire. Two hundred Balubas were there at first. About 300 more appeared after, so there were 500 in all.

Then Gleeson told us to load our weapons, so we loaded them. When the arrows had come at us we got the order to fire. Gleeson had no ammunition, and Fennell, Gaynor and Farrell were not armed. Fitzpatrick was missing. Only four of us were armed: myself, [Peter] Kelly, Michael McGuinn and Anthony Browne. Gleeson called him [Browne] for a magazine. I had a rifle and so did McGuinn. Browne had a Gustav and Gleeson also had a Gustav.

Gleeson gave the order to return fire and we fired. The Balubas only had

12. Pte Tom Kenny is tended to by army medics following the Niemba ambush.
Note two Baluba arrows hanging from his buttocks

bows, arrows, spears and clubs. Then Gleeson turned white as a sheet. He said:
'We'll pray.' He began to pray. Then he gave the order: 'Get away. Get away
everyone.' I looked down the length of his body to see where he had been hit.
There was an arrow in his knee. The blood had gone to his lower legs and was
in his socks. He was wearing shorts.

Eleven blacks were killed. Thirty-three graves were found. Forty-six of them
could not be accounted for. Twelve Balubas were taken to hospital in our cars.

During the ambush, Browne ran past me. When the Balubas came following, they thought I was Browne. They beat me on the arms, legs and head with their clubs. To this day the bones in my right arm are set crooked. I had an arrow in my neck. They put their feet on my face and pulled out the arrow. I think part of the arrow tip was caught in my shirt collar. I felt my feet go numb and the pain spread up along my body. I had two arrows in my buttocks. I wanted to be a pretty corpse so I put my hand up to protect my face. They continued to beat me. I was beaten to a pulp. They thought I was Browne.

Later I heard a bird singing up in a tree, so I thought that perhaps they'd all gone. There was blood coming from the back of my head. And blood from my arm had flowed into a pool and was like jelly. I felt like a smoke and reached into my pocket to get the packet of Belga cigarettes there. Then I reached down to my trouser pocket for the matches but they would not light because they were damp. Then I smelled smoke from a fire. I think the Balubas had set fire to part of the forest to try and kill me. I got up and started walking through the forest. As a youth I had shot pheasants in Wicklow and had fished in the rivers. These thoughts kept me going; I thought I was out shooting again.

As I walked through the forest I tried to take the arrows out of my behind. I tucked one of them into my boot and it snapped. The arrow in my neck could have killed me but the barb – which I later learned had been coated in snake venom – got caught in my shirt collar.

The patrols were out looking for us the day after the massacre, but even though there were 150 of them they kept to the road and did not search the jungle. They should have searched for men in the forest but they did not. When darkness fell they got in the vehicles and went back to base. They recommenced the search the day after, which was when they found me. On the day of the ambush I weighed 10st 6lbs, but two days later I was only 7st 3lb. When I was found they airlifted me by helicopter to hospital in Albertville. A member of the helicopter crew gave me a packet of cigarettes and some matches, so I finally got a smoke. I stayed in hospital until Christmas.

Whatever about the atrocities, I feel sorry for the black people because they were badly treated in the colonial days. I will never condemn the Congolese people, even after what happened to me at Niemba. They are lovely people.

I was back home in Dublin in February 1961, based in Clancy Barracks. I continued to be treated for my injuries at St Bricin's Hospital. I could not move my right hand. The swelling in my arms had gone down but the bones were still broken. They let me out of the hospital after twenty-eight days but I was put back in again. The treatment continued for about five months. When I was in the barracks I would go out on working parties. I served almost five years in the Army. They wanted me to stay but I could not work with the Engineering Corps because I could not move the fingers of my right hand properly. I was eventually released on medical grounds.

One day when I was at home, they sent a car to bring me to a court of inquiry at Army Headquarters on Infirmary Road. Five officers were sitting there with my eight-page statement [on the Niemba ambush]. They were Dan Crowley and P. D.

Hogan, who were sitting together, and Lt Col Gray.[4] There were two others I did not know. They kept asking me questions. They would ask me about something on page seven of my statement. They were trying to find out the truth. Crowley and Hogan were friendly. The others were there to stick the knife in and turn it – the sort that would say: 'Take this f—r out and shoot him. We gave him a fair trial.'

On another occasion, I was brought to Collins Barracks. They wanted me to present a medal to Anthony Browne's parents. I had developed fallen arches in my feet and could not walk properly. When I got there I started to protest. I was saying: 'Hold on, don't do it.' Nobody would listen to me. They carried me bodily out the door. They went ahead [with the ceremony] without me. I did not see Browne's family.

Eleven men went out on patrol that day. Three survived the massacre, yet nine coffins came home. The Minister for Defence, Mr. Michael Smith, referred to the eight victims of Niemba. Browne did not die in the ambush. He was very lucky at cards, playing poker. When he emerged from the forest he produced a roll of 20,000 francs to pay some Baluba women for food. When they saw the money they wanted it. They went to get food but came back with some local men who beat Browne to death.

Hogan should have withdrawn his statement that Browne died saving my life, because it is not true. If it can be shown that I stated that in the eight-page statement I made while in hospital in Albertville then I would accept it, but I did not state that. It did not happen. My eight-page statement has disappeared, but I wish someone could find it in the UN or elsewhere. When I asked the Army for a copy of my statement I got an unsigned three-page one which is not mine.[5]

My children and grandchildren have been told that Browne laid down his life to save mine, but it is not true. I don't know why this story was told. Only recently I was asked by some Army people if I wanted to take a medal off a dead man, but I don't.

I do not consider myself to be a hero. Lots of people go out on peacekeeping missions but they are not necessarily heroes. I am not bitter but I grieve when each new consignment of soldiers goes out to foreign service with the UN. I hope we will never have to erect any more memorials to them. It is nice to think that the soldiers of our small nation can be accepted as peacekeepers. We can hold our heads up anywhere we go.

I recently told the Minister for Defence that I could die happy because after making the mistake of giving Browne a medal, they came up with the idea of giving the eight who died in the ambush medals for the ultimate price they paid with their lives. It was 38 years too late, but I was not going to allow one or two to get medals. I still see the guys every day of the week; they are still young men to me. Somebody went out to the Congo, but a different person came back. I've never been able to find the person who went.

11

A Niemba Survivor Remembers

Joseph Fitzpatrick grew up in Dublin and joined the FCA (army reserve) when he was only 16 years of age. Here he recounts the horrific events of the massacre at Niemba on 8 November 1960.

I joined the Army in 1959 and went to the Congo in 1960. I was based in Cathal Brugha barracks in Rathmines. We did ordinary training for the Congo, although we didn't know anything about it. When we got the chance to go to the Congo, Jaysus – if anyone in Dublin got into the Sunshine Home you were lucky! Here we were going to the other part of the world. I looked on it as a holiday, to get to the other side of the world, that's all. I didn't realise at the time that I was heading into a war. When we eventually got onto the planes, there were people from Donegal, Cork, Wexford and Wicklow. You didn't know who the fellow sitting beside you was, so you were in limbo as regards companionship. I did not know any of the other fellows, they were just picked out of a group.

The first battalion to go out was the 32nd. I wanted to go on that but I was rejected for some reason or other. When the 33rd Battalion was formed I was rejected then also. But some fellow's mother didn't want him to go at the last minute so they let me go in his place.

It took two months for them to get proper clothing for us in Africa. Before that we wore the bull's wool shirts; we were going round sweating like pigs, but in the end we ended up with tropical gear. The battalion commander was Colonel Bunworth, but I didn't know him from Adam. You would never have a chance to talk to him unless you had committed a crime and were up in front of him. That would be the only chance you'd have of meeting him.

When we went to the Congo we landed in Kamina where there was a big air base. They had two-storey houses and we used to do parades around the airfield. We mounted fire pickets also. We were on duty all the time so we didn't get bored. The 33rd Battalion then went on to Albertville, while a small number – ten or twelve people – stayed on at Kamina. Every month we were called up to Albertville to join the 33rd Battalion. When we got there they had gone on to Niemba. Around the 6th November 1960 we were going up to Albertville and the tribesmen were burning the villages on the way in to Albertville. They were going to destroy the place. The Irish troops had dug foxholes and mounted machine guns, but they never came. The next day, Monday, we went to board the train. But it was like a bad omen because the locomotive wouldn't start. At about 12 o'clock we got started and went on our way to Niemba. We stopped on the way and had a cup of tea. It was a really eerie place, like a ravine with two high hills and trees on top. We went on to Niemba then and Sgt Gaynor picked us up in a Volkswagen and a pick-up truck. They brought us up into the hills at Niemba. On the way up there were all these mud huts, like our cottages here at home with the weeds coming through the roof. We were introduced to the people at Niemba. It was just a little town on top of a hill surrounded by villages. We settled into a place that was like a morgue. We had our mosquito nets and used to sleep on stretchers.

The next day, Lt Gleeson said we would go out and see the villages, to see the people who were hostile. Jaysus, we only arrived there on Monday, ten of us altogether. On Tuesday we went out on patrol but we hadn't got a clue about the terrain or anything else. We had a fellow with us called Corporal Duggan who had served previously with the British army in Malaysia. We felt safe with him being there with us. We went along and came to a big crater in the road, but they put these railway sleepers across the road so we could drive across. Someone came out of a village and started waving but we didn't know what the idea was, really. We went down the road and came to another big crater in the road. There were some of these tribesmen out in the road and Lt. Gleeson stopped the pick-up truck and the jeep. He went down to investigate the bridge and he put me and Browne on guard at the back of the truck. There were Balubas moving around in the bush, but about ten minutes later a big crowd – I suppose, ten abreast – got out on the road and started screaming and roaring. I called for Gleeson to come back up. He must have been about 100 yards down at the other side of the bridge, so they all had to fly up. Of course, there was big terrain at both sides of the road which was only a dirt track. Only something the size of a jeep could get along it. We formed a line and Gleeson told us not to fire. He had orders from the UN. We were on police duty, more or less, and we weren't to fire until they fired first. Someone fired anyway and that started the firing then and we just had to withdraw into the bush. We didn't know where we were going. Someone fired a shot. It could possibly have been one of ours, who knows? The other people had rifles as well so it was hard to know who shot first. A shot went off anyway and we had to withdraw into the bush. Gleeson covered us on the road until we got into the bush. I think he gave orders for someone to turn the trucks, but I don't think they could turn them because there was a big crater down the road and they had a tree across the bridge, and they had dropped a tree at the other end to block us in. Gleeson was firing the Gustav at this time. We withdrew and crossed the river. Browne and Fennel, who were small men, slipped off the riverbank. It was hilarious. The arrows were going over and all you could think about was Tom Nix and Roy Rogers back in Camden Street, going to the pictures. Jaysus, it was really happening. We got to the other side and met this chap, Gerry Killeen, who was a cook. He only went out that particular day to relieve another soldier. He was a cook and he wanted a day out. He was white and blue in the face and the sweat was just pouring out of him like a tap. He said to me: 'Fitzie, we're all going to be killed.' I went on into the bush and I reached this hill. At that time I didn't know what to do. The old army motto is: 'Yours is not to reason why, yours is just to do or die.' You couldn't even use your own brain to know what to do. You actually had to be told what to bloody well do, you were only like a child.

We got to a certain hill and there were about five soldiers on it. The tribesmen came round [outnumbering us by] about ten to one. Gleeson got an arrow in the knee and he got an arrow through his arm. He said: 'Take cover lads, we're all going to be killed.' So I went on into the bush then. A fellow followed me. I had a pair of boots on me and short trousers. I turned around and fired from the hip. I had a .303 rifle which I had to reload every time I fired. I shot him, he went backwards. He had something in his hand like a hatchet. I went on into the swamp. I met Browne then. He was spraying his Gustav. I met another fellow, Farrell, who was running. He had short trousers too, he was only about eighteen. He said: 'Fitzie, I'm after being hit by a poisoned arrow.' I don't know whether the arrow went through his short trousers or through his leg, but once you got hit by a poisoned arrow you'd only three minutes to live. It's the venom of the snake. I went through these green water lilies and knelt down at the other side in a firing position. I put the bullets in my shirt pockets to have them handy. I heard an awful lot of cursing and swearing. These fellows were going around with bows and arrows. They were like Apaches, they had bands around their heads. I was a while in the bush and then a big crowd got behind the bush that I was in and the diarrhoea ran out of me. The sweat dropped from my fingernails. They all roared like at a football match and they moved off. There was another chap beside me, Gerry Killeen, who had got an arrow in his shoulder. I tried to pull the bloody thing out but it was like a fishhook. Then I gave him my bayonet, I don't know why but it was like a John Wayne film, thinking you could cut it out. So he asked me to kill him then and I couldn't. I said: 'I can't do that Gerry.' I think he died a few minutes later. I just fumbled through my pockets. I had a cross and a prayer book. I fumbled through the pages and the Requiem Mass came up. I said a few prayers. Then another coloured man came in and I shot him. He was moaning and I shot him again. I was there all night in the bush. There was no twilight, it just got dark all of a sudden. There were big fireflies flying around. Lovely. Then the monsoon came. The trees were crackling and of course you thought they were coming in to get you, but they didn't really come. I stayed there all night. I said to myself if I go asleep now I'll wake up with no head. These were the ideas that were going through my mind. The next morning I heard a cock crowing. I thought I'd be there for three days so I blackened my face like I did in the FCA years earlier. A big snake came down the branch, but if I had fired a shot I would have alerted the jungle. I just hit it with the butt of my rifle and got out of there. I decided to find a river, but I didn't want to go back the way I had come. There

again, I didn't know where I was. I was in the jungle. I crawled over mercenaries on the ground with bandoliers on their chests. They had been killed.

There was a fellow called Farrell, a medical orderly. He was a knight in
shining armour, looking after people's toes and giving them tablets for a
headache. He had no weapon. His head was bashed into the ground like a
turnip. I went on then, in through this thorn bush which got around my neck so
I pulled it away. I didn't feel anything because the adrenalin was running very
high at that time. I decided to go across the river and went into a hole so the
bloody rifle got full of water. I didn't know if it could fire or not. I couldn't take
a chance on trying it. I got to the other side of the river and there was no cover.
The trees were all like sticks so I knelt down at this wall and I heard a sound
like oil drums falling. At that stage I didn't care who I was going to bump into.
I went down the road and it was the Ethiopians with the UN. They were more
frightened than I was because they didn't know what happened. All they saw
were red cartridges all over the road and blood everywhere.

I was not hit by a poisoned arrow but Tom Kenny was, although I did not see
him. His whole arm was like a big German salami sausage roll. His head was
like a turnip cut off at the top. It was all exposed with blood. He got two arrows
in his back and an arrow in his neck. I don't know how the poison didn't kill him.
Once you got a belt of one of these arrows you knew, 99 times out of 100, they
were poisoned, so you killed yourself. But you'd only got a few minutes to live.

Our patrol was going out from Niemba to Manono to keep the road open for
the Conakat[5] which, I think, was President Tshombe's special guard force.
Someone said that the black people thought we were Belgians, and that the
Belgians had been going in for centuries, raping their children, killing their folk
and destroying their villages. I don't know. Who knows?

I never even saw the bridge, I just saw the crater in the road. We were on
guard at the back of the truck and they went down to investigate it. Thomas
Kenny was in the Engineers so he would have known all about bridges. Lt
Gleeson would have known about bridges as well.

We got out of the jeep and the pick-up truck. Lt Gleeson and Sgt. Gaynor
got out. They saw a number of these tribesmen on the road. One fellow had a
leopard-skin front and hat. Gleeson said 'Jambo. We greet you in peace' and
they withdrew into the bush. Then Gleeson, Gaynor and a couple of other soldiers went down across the bridge to investigate. Within ten minutes – you
wouldn't have time to lift a shovel – they had to withdraw. I roared for them to
come back up to meet the tribesmen. Part of the patrol's duty was to repair the
bridge and keep the road open. We had only been there for fifteen minutes – just
time to survey the area and see what had to be done to repair it – when it all
happened. They didn't even get time, it was only a thought in their mind. Of
course, the Balubas wanted to keep that road closed.

As regards whether anything could have been done to better protect the
patrol, first, they should have had an interpreter with them, instead of which
you had people speaking English to a crowd of people who didn't understand

the language. We couldn't communicate with them. Also, as regards myself and a couple of the others, it was our first day out there. We didn't know the people were hostile, we didn't know what they were like. We were only going from point A to point B, yet we were in a mess because of it. If we had been familiar with the terrain things could have been different. But as a soldier you were not allowed to use your own mind, you had to be told what to do. The Balubas outnumbered us by about ten to one, although we had more guns. I don't think they got the bren gun out of the car at the time. There was a UN order that the bren gun wasn't to be used. It would have been genocide to the black people. Our job was only to do police duty and not to fire unless we were fired at first. The bren gun could fire about 100 rounds per minute and its use would have resulted in genocide. The gustav was a small hand machine-gun. I was a rifle man so I didn't know about the other weapons. Every time I fired a shot I had to reload. It was a terrible situation. I might as well have been Billy the Kid or John Wayne. After the crisis at Niemba they brought in automatic weapons.

I do not know anything about a Baluba chief's son being shot about two weeks before the ambush at Niemba. We only arrived there the day before, but people had been based at Niemba for a month or two prior to our arrival, so they might have known. We arrived there on a Monday and were told that the following day we were going out on patrol into the bush country. I didn't know anything about it. The Balubas may have thought we were Belgians and, as I said, the Belgians had crucified those people for God knows how long, raping their children, killing their fathers and robbing their sons. Someone said they mistook us for Belgians, but I don't know. I'm not going to swear on it. All the UN soldiers wore blue helmets, but I don't remember having one on myself that day, now that I think about it. Maybe they took them off, I can't say. It is forty years ago.

Three hundred thousand people turned out for the funerals in Dublin. Everyone had rosary beads in their hands. We were in hospital in Africa. It was the biggest funeral procession since those of President Douglas Hyde and Parnell. With such a big turn out you would imagine that the men would have been honoured but they were not. It took thirty-eight years, but myself and Kenny have not been honoured. Recently a pigeon was awarded a medal for bringing the news to the Allies that the war was over. According to that, a pigeon is more important than me and Kenny, the two survivors. We will have to wait until we die to get a medal of honour on our coffins. So we live in shame but will die with honour. The Army reckons that you have to die to be honoured. At a later stage in the Congo, an Irish Army soldier received a medal of honour for carrying a breakfast of rashers and sausages under fire. The Taoiseach, Bertie Ahern, is aware of the situation. I have sent him a copy of my book [My Time in the Congo, pamphlet privately published 1998]. The Minister for Defence is also aware of it. The UN, television and the newspapers are aware of it, as well as the President of the United States, so there is nothing more I can do. I honestly cannot see anything happening at this stage. I am just happy that the lads got fixed up after thirty-eight years. My good name was taken away from me. People said I ran away. I just had to live with it. People said: 'You were a coward, you ran away.' Neither I nor Kenny ran away. How could we have done? We had a purpose in life: to do or die, not to ask the reason why. The officer told us to take cover or we would all be killed, and we obeyed orders.

It is possible that the Army made a mistake. There was a young man there called Davis. His father was in the Irish Army during the Emergency and his brother was in the SAS in England. He wanted to join the Army when he was sixteen but he couldn't get in. So he used the birth certificate of his brother who had died at birth. His name was Willie Davis but everyone called him Paddy. He was shot accidentally in Niemba, by one of our own men – shot through a door. So maybe the Army didn't want to bring out their dirty linen in public. I don't know. After the nine men were killed at Niemba he was killed by one of our own men at the outpost on the hill in Niemba village. The tribesmen were attacking the outpost and he was shot dead. It was an accident. He is mentioned now as the tenth man who died at Niemba. His name is inscribed on the roll of honour. What can I say? Are they inhuman, or what?

13. Private Joseph Fitzpatrick at the Kamina Air Base, Congo, 1960

12

A Swedish Interpreter Remembers

*Swedish Army veteran, Stig von Bayer, served in the newly independ-
ent Congo with the UN's peacekeeping forces. He acted as an
interpreter and liaison officer for the Irish Army during the fraught
period of the Niemba ambush in November 1960.*

My father and mother were involved in building roads and bridges in Kivu
Province from 1948 to 1960, as well as running a palm oil plantation. I spent
six years with them there and that's how I learned to speak Swahili and French.
I arrived there in the first half of 1949 and stayed with my parents the first two
years, thus learning perfect Swahili, going out hunting on a daily basis and
feeding the workers in this way. Then I went to the Athénée Royal boarding
school in Bukavu, remaining at the school and hunting on vacations for the
workers, until my return to Sweden in 1956.

My first contact with the Irish Army came about just after I arrived in
Léopoldville at the end of July 1960. I came directly from my regiment, where
I was a platoon commander. I only stayed a few days with the Swedish battalion in Léopoldville before going with the [Irish] 32nd Battalion to Goma and
later to the 33rd Battalion in Albertville, where I remained until January 1961.
I was stationed with the 8th Swedish Battalion and had been there for three days
when they woke me up at two o'clock in the morning to tell me that I was going
to Goma. 'Excellent,' I replied, 'as my parents are missing somewhere in that
area.' 'You are going as an interpreter and liaison officer to the 32nd Irish
Battalion,' they said. 'Impossible,' I replied, 'as I don't speak English.'
'Bullshit,' they said, adding 'and hurry up.'

The first Irish person I met fitted exactly my mental picture of a typical
Irishman – in this case, a doctor – stirring eyes, a stick and speaking a version
of English I didn't understand a word of. Fortunately, Comdt Joe Adams then
arrived. He spoke perhaps the best English I had ever heard; I understood at
least half of it! 'We are going in on the first plane to Goma, today,' he told me.
When we arrived in Goma the airfield was completely surrounded by a muti-
nous ANC [armée nationale congolaise or Congolese national army] battalion,
ready to open fire on us.

Together with Comdt Adams, we spoke to the ANC company's commanding officer. I understood perfectly well what those idiots wanted, but had great

problems translating it into English. I told them we were not Belgian para-troopers, but that we were coming in to help them with all their problems, espe-cially the Belgian paratroopers. In those days, I spoke better rural Swahili than practically any Belgian could. So, in the end, they believed us and put down their cannons, mortars and medium machineguns (mmgs), letting us unload and occupy an old school building. We wouldn't have had a chance if they had start-ed firing at us as all the Irish soldiers were still in the planes – C124 US MATS (military air transportation).

I remained in Goma for about two weeks doing patrols in, among other places, the National Park where the ANC were running around using the ani-mals for target practice with mortars, mmgs and small arms (pistols and rifles). We had some problems in convincing them that this was not the right thing to do.

When I joined the 33rd Battalion in Albertville, they had just received the 84mm Carl-Gustav recoilless guns but had not yet had a chance to use them. What a coincidence! My last job with my regiment in Sweden was as the pla-toon commander of an 84mm CG unit, so we immediately set out to train the Irish lads with their new weapons. Not having any exercise ammunition we were only using live ammo, namely high explosive (HE) and anti-tank, at the local firing range at 300 metres' distance. The first HE shot convinced every-body to remain behind a protective wall until well afterwards, thus learning the tough Swedish security rules the other way around. As far as I know, the Irish troops never had any serious incident with that weapon, although the Swedish battalion had several such incidents during the fighting. (While we were shoot-ing with the 84mm Gustav weapons, the high explosive ammunition detonated on impact or after a certain distance, thus spreading shrapnel in all directions. Even at a distance of 300 metres from the target, pieces of shrapnel came whizzing by, forcing everyone to lie down. The anti-tank ammunition could burn a hole right through the thickest armour. It could be fired from the 84mm Gustav at a reasonable distance, while standing. The Swedish UN soldiers fre-quently fired the high explosive ammunition but several were hit by shrapnel from their own weapons.)

I remember the events of 7th to 10th November 1960 very well. I was with P.D. Hogan's patrol (all of them) on the 7th of November, and Lt Gleeson was also with us. It took us the whole day to reach the bridge, with all small bridges destroyed, trees blocking the road and elephant-traps – the head jeep actually disappeared into one. Afterwards, we had to be very careful not to fall into the next one, a big hole fully covered with sand on top, although the first one was not dangerously deep. Militarily we were on full alert with patrols scanning around during each stop, and sentries, thus presenting a picture to the Balubas who were hiding in the surrounding area and running away when approached by our patrols. Upon reaching the bridge we realized that it would take one or two days' work to fix it and Gleeson got the order to patrol on a daily basis in order to keep this stretch of road open.

I spoke to Lt Gleeson after he got the order from Lt Col P. D. Hogan to patrol the bridge at Niemba on a daily basis, and I said he should be very careful with the Balubas. I warned him that they could be extremely dangerous, especially when taking drugs. He answered, 'No problem. They know that we are the UN and are here to help the Balubas.' I could only repeat, 'Don't believe that!' Gleeson, who so far had encountered only kind-natured Balubas around Niemba, merely laughed, saying that they knew we were their friends and were there to help them. I repeated the warning strongly, asking him to be extremely careful. Gleeson answered, 'No problem!' I think I repeated the conversation in general terms to P. D. Hogan. However, Gleeson was ordered to have an mmg mounted on one of his vehicles.

After many patrols together with P. D. Hogan and close encounters with the Balubas, I was considered the general Congo combat expert. I acted as a staff officer with the team. The company sergeant-major, with his sergeants, liked to have me around in order to decide on defence perimeters and patrols, including speaking with and instructing the soldiers. We developed a tough, no-nonsense attitude, which created respect. You could always see the Balubas hovering around at a safe distance. One person who, I remember, warmly supported this tough stance was Lt Raftery, the engineering officer. It must also be noted that our attitude was quite different from that of Gleeson's platoon, which adopted a more peaceable, civilian attitude. This was drastically changed after the attack.

Lt Gleeson had an order to patrol daily up to the bridge and nothing else. Upon arrival at the bridge instead of turning the vehicles around pointing in the direction of Niemba, all or most of the eleven-man patrol left the vehicles – some probably without weapons and spare ammo – and went over to the other side, thus giving more a picture of a picnic than a military patrol. Had they all remained around the vehicles with the mounted mmg and perhaps going over with a well-armed patrol of two men, they would no doubt have made it home again, providing they did not let the Balubas within machete range. It must be remembered that when we interrogated the Baluba prisoners they said the people in the patrol looked kind. I interpreted the Swahili term 'not dangerous' or 'inoffensive' as meaning 'easy prey' in this context. According to the Baluba prisoners the patrol that had been there the day before (7 November 1960) 'looked angry' and displayed a 'no-nonsense military behaviour'. It was, therefore, more important for the smaller party, of 8 November, to show an even higher military attitude of readiness. Gleeson was, in fact, warned about this but either didn't listen or did not believe it.

We were back in Albertville when we received the information that something terrible must have happened to Gleeson's patrol. We immediately organised a strong rescue operation. When we reached Niemba the day after the ambush (9 November 1960), we were informed that a big Baluba war party had occupied the railway station on the other side of the valley. I suggested that we should drive them off for security reasons. This suggestion was supported by all

14. Stig von Bayer of the Swedish
Army in UN uniform

the lads but the Commanding Officer said we must have UN HQ's permission
and, of course, the reply was that you must parley. During the night heavy
shooting started around the compound. I went out of the building together with
Col P. D. Hogan and Comdt McMahon. We asked a soldier in front of us what
he was shooting at. In the meantime another soldier passed behind us in front
of a window. He was instantly killed by a soldier who had remained in the
house and thought this was an attacking Baluba. This man no doubt saved our
lives. Lesson learned: every man must have his position and exercise finding it
night and day. It was very unfortunate for morale that we let the Balubas be.

In a UN operation many questions can't be asked: we carried out a strong
patrol around Niemba; we were fired upon; fire was returned; casualties
unknown. Patrol now back in camp. There were no written orders but UN HQ
didn't want any body counts; they must be protected from these realities.

The UN orders were to fire only when fired upon. But when warriors encir-
cled UN soldiers and tried to take away their weapons within machete range, it
was considered that self-defence was justified. This happened several times to
the Swedes on the Luena-Bukama railway line as well as three or four shootouts
in the Baluba camp at Elisabethville.

Shortly after we left Niemba in the morning for the bridge, there was a new
shootout in the camp and one soldier who was entering a room got shot in the

stomach by a friend who was shooting out the window at some unseen Balubas. Fortunately, the helicopter with Colonel Burns (I think) arrived just then so it was possible to save the soldier's life.

We left Niemba on the morning of 10 November, bound for the ambush site, with a strong patrol reinforced by an Ethiopian unit. I was in the first jeep, when we heard a shout and saw somebody just to my right coming out of the bush. It was Private Kenny stumbling out with two arrows in his buttocks. When he saw P. D. Hogan in the second jeep he stood to attention and said: 'Private 57 Kenny reporting, sir.' He was immediately taken care of by the medics and when the Ethiopian doctor cut out the arrows Kenny was screaming. Soon afterwards, he was evacuated from the scene by helicopter. Private Fitzpatrick turned up later carrying his rifle and, as it seemed, unhurt. We carried on to the bridge finding some bodies but we never found Trooper Browne.

After some time in the Niemba region we learned about waging African warfare against so-called invulnerable warriors – that is, never to fire any warning shots in the air as this only confirmed their belief that all bullets turned into water and would be the signal for an immediate attack.

In those days, the Baluba warriors were all drug-crazed, taking strong medicine mixed with narcotics, which they thought made them invulnerable to bullets. The amazing truth was that when they were hit by a bullet they didn't feel anything and very little, if any, blood came out. The drug made them stark raving mad and, of course, it was absolutely useless to try and parley with these people.

In most cases, in the 1964–5 period, when the Congolese army fired at these bands called Simbas or Maj-maj, they just kept coming and the Congolese threw away their weapons and ran off. Of course, they died later but in many cases not before having killed their opponents. The same Maj-maj bands are still running amok today in Eastern Congo.

INTERROGATION OF BALUBA PRISONERS

My interrogations of the Baluba prisoners following the ambush were, to say the least, sensational. One of the Baluba prisoners indicated that the witchdoctor wanted to use some body parts for strong medicine but this was, as far as I know, never confirmed. They found Browne's body about two years later, I think, but an autopsy would probably have been impossible by then.

It is important to remember that the only one taking notes during the first interrogation at the spot was either P. D. Hogan or Captain Sloane. Thereafter, the Baluba prisoners were handed over to HQ and evacuated by helicopter. We never saw them again. We had, as prisoners, one elderly man and two or three kids. The old man spoke willingly, the others very little. From memory, the interrogation of the Balubas went as follows:

Q. Why did you attack the UN patrol?
A. The witchdoctor said that anybody coming along this road must die.

Q. Do you know what the UN is here for?
A. Yes, they are here to help us.

Q. Did you see the patrol the day before?
A. Yes, we did.

Q. Why didn't you attack us then?
A. You looked so angry. [Here you must note that many words in Swahili have more than one meaning. The best translation might be: You looked like a tough, no-nonsense military unit.]

Q. And the patrol next day?
A. They looked so kind. [Translated to: inoffensive or easy prey.]

LUMUMBA ASSASSINATION

As regards the assassination of Patrice Lumumba [on 17 January 1961], the Swedish soldiers did, in fact, first observe the arrival of three Africans [at Elisabethville airport] whom the Katangese were beating up on the tarmac, before being loaded into a pick-up truck. One of them was identified by a Swedish soldier as most probably being Lumumba, and this was later confirmed by several Katangese.

It is quite correct that there were Swedish troops on duty at the airport but they were under strict orders not to interfere with the Katangese traffic. As a matter of fact, one part of the airport was under Katangese control only and no UN personnel had access to this area. This was an essential part of the agreement between Dag Hammarskjöld and Moïse Tshombe, under which the UN was allowed into Katanga. This arrangement was changed when the UN got sufficient reinforcements in place there.

13

Plus Ça Change…

Lt Col Eoghan O'Neill commanded the 34th Battalion in the Congo, arriving there in January 1961. This is his story.

After being promoted to the rank of Lt Colonel, I was placed in charge of the Cadet School where I remained for three years. During that period I was sent to the Congo. I was nominated as a Battalion Commander – the next one after Col Dickie Bunworth's 33rd Battalion. The first battalion to go out was the 32nd. That number was picked because the 31st was the senior number in the Army during the Emergency. The 31st had been commanded by Col Mossy Donegan who rose from the rank of Private soldier to Lt Colonel within twelve months. He was second in command to Tom Barry in the West Cork brigade IRA during the War of Independence.

It so happened that the 34th Battalion was the next to go out to the Congo after Niemba. Niemba taught us a lot. We learned that you have to be ready for anything, even in the mildest of circumstances. We went out in January 1961 and I knew before Christmas that we were going. Just after Christmas 1960 it was decided that we would replace the 33rd Battalion. The 32nd Battalion was to return at virtually the same time but would not be replaced. Those two battalions – 32nd and 33rd – had been formed into an under-strength brigade. Normally, a brigade should have least three battalions. So the brigade staff were not replaced and they had gone by the time I got out.

We landed at Léopoldville and were then sent 300 miles or so to the centre of the Congo, a place called Kamina. The base there was a military installation as big as Aldershot and bigger than the Curragh camp. They say it was built originally for NATO and it was a training base for the Belgian air force. The weather was so perfect you could fly there 365 days a year. There was a big town about five miles long as well as civil quarters comprising private houses that had previously been used by Belgian officers. There was also a big civilian body there because it was the administrative centre for the region. There was a town within that called, unfortunately, the native town. There were houses for the Congolese. They were from different tribes but got on well together. It was Baluba country and we got on well with them. We were required to continue the administration so we had to put our men into a farm of about 250 acres. The crops were bananas and pineapples, and there was a herd of about 250 cattle which had to be looked after. There was also a market with a butchery to which we supplied cattle. A sergeant was in charge of it with a couple of men and a number of Congolese helpers.

We took over the nearby airport and found someone to be airport manager. There were 30 flights a day, which was about the same volume as Shannon airport at the time. We wanted somebody who could speak French and apart from our legal officer, the only person in the battalion who could speak the language was a corporal. He was put in charge and he ran the airport, although he never had any such experience before. He did a great job, as did the NCO in charge of the workshops. We had two transport platoons run by a Pakistani and an Indian NCO. We also had a police company consisting of three Scandinavian contingents – Danes, Norwegians and Swedes. We put someone in charge of that but the Scandinavians ran a native police company of about 300 men who were very well trained. It was largely a paramilitary force. Altogether, I had about 500 men under my command.

There were other bodies that were under our command for administrative purposes. We had only been there a few months when Lt General McKeown informed us that the King of Morocco had died. There was trouble in Morocco at the time over the succession and whether or not they would have a new king at all. We did not know much about it but the two Moroccan UN regiments in the Congo decided to return home and the Moroccan Government did not want this to happen. The two regiments comprised elite troops; one was from the former French Morocco and had been part of the French army. They were Colonel Guillaume's famous Moroccan force, which had gone up along the backbone of Italy in 1944, taking Monte Cassino and other objectives. They were tough, but equally tough were the troops of the other regiments which came from the former Spanish Morocco. They had been part of General Franco's Moorish cavalry which captured Madrid in the Spanish civil war. Their punishments were very tough. The head of the Moroccan army in the Congo at that time was General Kettani who was second in command to Lt General McKeown. The Moroccan government may have been frightened of him returning home as he was considered to be the strong man of Morocco. The government did not want him to return at the time. In later years, I went on holidays to Morocco in a place called Agadir where the main street is named after General Kettani. He was honoured after his service in the Congo.

We were comfortable in the Congo, with plenty of lovely billets. We were able to put some of the officers in suburban civilian quarters. We had two wonderful messes – one for the officers and one for the men – as well as two Olympic-size swimming pools. It was marvellous. One of our officers became sports officer and every day there were games of hurling, Gaelic football and soccer which was not all that popular then in the army.

We had nobody to compete against in hurling. We did a lot of training, firing our range practices. We had to do such training as we had been brought together in a hurry and so had the two preceding battalions. It meant that we had an ad hoc battalion. The military method of organisation is that a commander is responsible for everything he commands – both his staff and his subordinate or company commanders. These in turn are responsible for everything in their units, where it concerns the administration of ordinary day-to-day living or action and operations. A commander should have trained his intelligence, operations and engineering officers. Fortunately in my battalion the officers were about the same age and had met one another either as cadets, on courses or having been stationed together. The lack of prior training, although wrong, was quite unavoidable. We were able to overcome it and got time to train. We were not in action but we had to do a number of patrols into what many would consider to be hostile country – Baluba territory. But the emperor of the Balubas lived near us and I called on him one day with an Indian officer. He was a very interesting chap. He had a jail of his own and he was a tough man. His section of the Balubas had nothing to do with the Balubas who attacked Irish troops at Niemba. It was a different territory and they did not recognise our Balubas. The emperor did not say anything about what had happened at Niemba nor did I mention it to him at all.

We had to exercise the men, firstly at section level – the smallest group commanded by a Corporal, then at platoon level commanded by a Lieutenant. There are three sections in a platoon and three platoons in a company. We ran company exercises with the battalion. We were also responsible for security but we got plenty of time for such exercises. We were surrounded by bush which was mostly tall pampas grass, ten or twelve feet high with paths through it. It was very different from operating through Irish countryside so we had to work out systems for providing covering fire and supporting fire. We trained men to move through the bush.

We also had one outpost at a big hydroelectric scheme which was supplying half the area. We had a lot of administrative work there too. We were getting on very well with the locals and coming up to Easter we had a very big day for St Patrick's Day. We had some thirty nationalities dining with us that weekend. They came from all over the Congo. Two Irish people came down for the party, including a man from Portadown, of that persuasion, and we got on very well.

Over the Easter weekend the Katangese rebel government, under Tshombe, seized the airport in Elisabethville, the capital of Katanga. Katanga was a terribly wealthy state which broke away from the Congo at the time of independence. I went into a private house there and I saw a lump of raw malachite in the hall. It is a semi-precious stone used in earrings and other jewellery. It was so big it was being used as a table. The wealth of that city was unbelievable. There was a coal mine within the city limits, an iron mine three miles away and gold mines thirty miles out. About the same distance away there was a uranium mine which had provided uranium for the first atomic bomb. The wealth of that! And then they had a number of very well-trained Belgians mostly in senior positions doing technical jobs. The Katangese gendarmerie was the local army commanded by Belgian officers and there was a big mercenary force – the cut-throats from all over Europe, including at least one Irishman. We did not have anything to do with them. There was a big demonstration in Elisabethville and some of the leaders got up and said they would march on the airport and take it over. Some 3,000 people went on the march led by a blonde buxom white Belgian woman who was shrieking. I got word from Lt General McKeown that we were to go in at first light and take over the airport. The following morning, Easter Monday, we made all our arrangements and got six or seven aircraft ready. A-Company from the Eastern Command went down first, under the command of Comdt Ned Vaughan. They were followed by B-Company from the Southern Command, and Comdt McDyer from the Curragh followed in after. We arrived there and they were not expecting us. We took over their sentries in the same way as the Belgians and the Katangese gendarmerie had taken over the Swedish sentries. Without firing a shot we occupied the airport. Lt General McKeown flew over it just before we came, and they threatened to shoot him down but they did not. Shortly afterwards they had to announce to him that they did not own the place anymore. We were very lucky in that we had trained for that kind of operation. The first company had taken over all the important posts. We had worked it out on paper before we went in so everybody knew where to go. They started digging straight away.

We were not very popular with anybody at that time, and certainly not with the Swedes for they had lost the airport. We were afraid there would be a counter-attack which is why we dug in. I remember sleeping in a very large stores tent with three or four others. We had slit trenches dug into it. The Belgians tried all sorts of silly – although I suppose they weren't silly – things to over-awe us. I remember a man coming up to me one day and he said: 'Look at this,' showing me a small bow and arrow, 'those are the things you fellows will be up against.' This was an attempt to overawe me. A few days later the airport manager came to tell me that we would be overrun in no time. I asked: 'Who will overrun us?' He replied: 'The civilians and the gendarmerie.' I asked him how they would do so and he gave me some spiel to break our morale. I said to him: 'By the way, I'd like to show you something here.' I took him up to the top of a building from where we could see the airport control tower where this Belgian lived and worked. I showed where we had deployed four heavy mortars, and I added: 'They are trained on your building, so no matter who comes in here you are going to go first.' He turned white. He pleaded with me not to give my men whiskey, so I asked why. He said: 'In Belgium [in 1944], I remember an Irish regiment in the British army coming in. They drank the place dry and nearly carved the place up.' I said: 'If you behave yourself we will see that nothing happens.' They did behave themselves but among other things they refused us permission to use the airport toilets. Elisabethville airport had the best restaurant in the city – a bit like Dublin airport when it first opened up. There was a lovely lawn there and many of the Belgian wives came out for afternoon tea. It was the thing to do. On a Sunday the big jet came in from Brussels with fresh lobster and other food, so the place to eat was the airport restaurant. They told us that they would allow the officers to use one of the airport toilets, but not the men. I said that was no trouble to us because we have a system for that. So we dug our open trench latrines on the lawn in the tea garden and there was nothing they could do about it. I do not want to sound anti-Belgian; I was merely against the people there who were trying to demoralise us in every way. We would not allow our soldiers to be treated in that way. Our reaction demoralised the Belgians more. We then began training exercises all around the airport with armoured cars and troops engaging in mock battles. That did a lot of good for our morale. As well as Belgians there were also Portuguese there from East Africa, and British.

We remained at the airport until we were relieved by the next battalion. We did not have to fire a shot. Our officers and men were not aggressive but we learned many lessons from what happened. We had to make them understand that we knew our job and that if the job had to be done we would have no compunction in shooting. They learned that lesson very early. We were extraordinarily lucky. The second battalion after us had to land at the airport under fire. Another battalion which went back up to Kamina, where we had been based, were strafed from the air by machine guns of the Katangese gendarmerie which had a couple of jet fighters. The third battalion after us had to fight its way back into Elisabethville in the battle of the tunnel. Comdt Feehily, who had been a cadet when I was in charge of the Cadet School, won the DSO medal for valour on his first journey out. His father had been my brigade commander when I was in Limerick. He is now a Colonel.

At Kamina we had a battalion of Gurkhas from the Indian army under our command. They were very interesting people to work with, and very good. We decided to amalgamate the Irish, Indian and Pakistani transport mechanics in one location where we set up a workshop. After a few days I went down to the workshop to find a soldier using a hurley to strike stones against the galvanised iron building. When he turned around I realised that he was a Gurkha who had learned hurling from our troops.

The transport personnel got on very well together. They included Muslims, Hindus and Christians but there were never any religious differences. The officer commanding transport told me one day that the men in the workshops wanted to commence fasting for the Muslim holy period of Ramadan. None of the Pakistanis or Moroccans could eat between sunrise and sunset. The officer explained that the Irish lads wanted to do the same, so we arranged that no food would be delivered there until 6 p.m. – the tropical sunset is very sudden – and again at 6 a.m. They lasted all day without any food. The Irish troops did that because they did not want to upset the others.

We learned an awful lot from the first two battalions that went to the Congo. Many of the officers overlapped with us and we got to know things that they had learned the hard way. We had been told that one should not upset the Africans because they might start playing rough. We were told to go easy with them, but our own people told us not to believe that because the Africans were playing a game. We were told never to forget we were a military force. We were encouraged to help the Africans in every way and put on concerts for them. We ran a kind of Feis with Congolese and Irish dancers. We were told, however, that no matter how the Africans treated us, we must always have a guard ready. We did so and the guards were always conspicuous at these functions, and that helped us a lot.

The first battalion to go out had a chaplain called Fr Paddy Crean who had been a chaplain in the Second World War with the British parachute regiment. He was a marvellous chaplain. He was head chaplain with the Defence Forces and he passed on a lot of good information to us. He ran the bingo games and also arranged recitations of the Rosary for the men. Some people laughed at that, saying that soldiers would not be so inclined, but that is not true. Every night when the Angelus sounded, you heard the Rosary being said. The Europeans there could not get over this because none of those practised at all, even though they had a bishop there. Our men were extremely good in that way. The way they would say the Rosary without fail certainly made me hold them in colossal respect. We had two chaplains and Mass was celebrated in every post. We had three or four posts and Mass was said every morning at 6 o'clock. On weekdays about one third of the men attended Mass, while others would be on duty. I suppose they had every reason to attend Mass, as it was a jittery situation out there. They were reading things in the newspapers and hearing the radio reports from Rhodesia, which were directed at UN troops to break their morale by telling what was going to happen. Rhodesia was very much in favour of the breakaway government in Katanga. The radio would claim that various tribes were arming up in such and such a place and intended to march on Elisabethville. Fortunately we were aware of what it was – propaganda to break the morale of the United Nations. It was meant, not to be instantly terrifying, but to build up a propaganda which gave you the idea that worldwide they were all against you. But my men knew they could handle the situation.

I met Dr Conor Cruise O'Brien there when we went down to Elisabethville. He was sent down there to relieve his predecessor as UN civil administrator, who was a Frenchman. The military leader for most of that time was a Swede, General von Horn. Lt General McKeown nominated me as his deputy. When the Swedish general went on leave for a month I took over. I met Dr O'Brien every morning. There was no great difficulty because nothing much was happening. We were very lucky, although some people would say we were unfortunate that we did not have action. Our officers felt, however, that while they were prepared to fight they would prefer not to have to. We suffered no casualties in action.

I found Dr O'Brien to be a very pleasant and witty man. While he is very intellectual he can move in any company. He also had very strong opinions and you will not change his mind very easily, not that we ever had any reason to. I had a daily briefing with him for about a fortnight, during which time I briefed him on the military situation telling him what our intelligence had heard. He assessed the situation from other information sources also and made a report to the UN. As to who was in overall charge, I would say there would have been a discussion. As to the exact relationship between us, if he asked me to use troops I would certainly have contacted Lt General McKeown first. I would accept my orders from McKeown. That was the way with Dr O'Brien's predecessor and he did not go beyond that with me, although I don't know what he did with my successors. That situation never really arose. Had he said: 'I want a company tomorrow,' I would be on the radio to Lt General McKeown straight away to see what was to be done. He would probably have said yes, but even if I agreed with Dr O'Brien I would have sought the authority of Lt General McKeown first because the military line was the one I would go by.

Katanga could never have been independent in the true sense of the word. It would have been run by Belgian and other international financiers. The biggest uranium mine in the world is located there. I had a villa there, which I didn't use very much and there was a gravel driveway. I picked up the broken stones and could see they were copper, just as you could see the same thing in various parts of Wicklow from the times of the copper mines there. I brought the copper pieces to show somebody who said that if they were found in Ireland they would re-open the mines, yet the waste from the copper mines in Katanga was being used as gravel. They were digging up solid pieces of malachite which was 49 per cent copper ore. The mineral production was controlled by an international mining company called Union Minière, which really owned Katanga and ran the puppet government. The international community was against them, however. If Katanga had succeeded in breaking away from the Congo and the United Nations had withdrawn, the rebel province could have succeeded financially. Militarily, it would have been able to put down a rebellion within its borders. They would have used force and would have been supported immediately by Belgium and Rhodesia.

Going into the Congo in 1961 we were not apprehensive, despite what had occurred in Niemba. If we knew we were secure we could not be apprehensive. We had done all sorts of drills beforehand. I remember at the square in Cathal Brugha barracks we did drills the night before we left: how to embark on and disembark from aircraft under fire. All we were doing was getting in and out of lorries but when Ned Vaughan went down to take Elisabethville, he did exactly the same drill. Without giving any orders, No. 1 platoon had to go forward 300 yards, No. 2 platoon went 300 yards to the left, and No. 3 went to the right. The platoons were spread out like that with the heavy covering force behind. That is how they took over the airport. The sentries were asleep; it was just before dawn. By daybreak we had control of the airport. There was no apprehension going in because we had trained so hard. The men were glad that the officers were checking up on them. When we went there first, very often the men had to sleep in bivouac shelters with loaded rifles beside them. I would be more apprehensive of a shot going off by accident in such cases.

We naturally had to control drinking and there were only certain hours during which troops could drink while off duty. There were only certain drinks you could have. The danger was that whiskey and gin was only five shillings a bottle. If they started on that, you have only to read any story about the tropics – particularly in Africa – to see what drinking gin or whiskey does to Europeans. They can't take it. Our troops were pretty abstemious, although I don't mean they did not necessarily drink the canteen dry on occasions.

The Irish presence in the Congo served a purpose at the time, but I am doubtful as to whether or not the purpose is still being served. Handing it over to Mobutu did not seem, in the long run, to serve any purpose either. He was as corrupt as anybody was. President Laurent Kabila was just the same.

14

Through Belgian Eyes:
The Decades of Transition

Henriette Cardon-Sips, from Antwerp, lived in the Congo from 1924 to 1967, and saw the transition from Belgian rule to independence at first hand. Here she looks back at the decades of transition with mixed feelings.

I was born in Antwerp, Belgium, on 4 February 1921. At the end of 1922 my father, Victor Sips, left for the Congo. He was a veteran of the First World War but had lived in the United States for four years before the war and could not get used to small Belgium after that experience, so he applied to work in the Congo as a territorial agent. He was too old to do a university course in colonial administration, so he left for the Congo. Because he was posted to a remote mountainous bush area in Katanga, they told my mother that she could not join him there. So she went nearly every day to the ministry of colonial affairs, and finally after sitting there for several months, one of the men said: 'Oh, for goodness sake, let her go.' So I left for the Congo with my mother on 4 February 1924 – my third birthday. The only thing I remember is that my grandmother cried. The journey, by sea and inland by train and riverboat, was so long that we did not meet up with my father until 4 May, three months later. Since my

15. Henriette Cardon-Sips (right) and her family take shelter behind sandbags in their Elisabethville home during a UN bombardment, December 1961

mother was the only unaccompanied woman on the three-week voyage from Antwerp to Matadi, she had to seek protection from the captain who posted a sailor outside our cabin door.

At the first port of call, which was La Rochelle, what my mother called 'les respectueuses' (the whores) got on board. They disembarked at Casablanca, where others got on board, and duly got off at Dakar. It continued like that until we arrived in Matadi. My father had instructed my mother to hire a black boy 'begging for work' when we got there to help us with the train and river trips, because 'they are the specialists and will do all the work'.

We got to Kinshasa, the capital city, just in time to learn that the riverboat had left, so we were stranded for several days. We always called the city Kinshasa, rather than Léopoldville, because my father did not accept the Belgian names. The only place with a Belgian name that he acknowledged was Albertville because he had fought in the war at Ypres for King Albert. He used to tell the story of King Albert's Christmas visit to the trenches. The soldiers were so fed up eating herring every day that they hung up the dried herrings as Christmas decorations for the monarch's visit. My father respected the Belgian monarchy but he thought it was right to use the African place-names. He did not understand why we had to give them Belgian names.

On the riverboat my mother had to ask for protection again because, apart from one couple, the other passengers were all single men. We ate goat's meat most days, but the smell of the goats was terrible – they were kept alive in the boat's hold. The boat made its way along the crocodile-infested Congo River but could not navigate at night. When we stopped at villages, the boat boys would run off to buy eggs. They tested the eggs in a bucket of water to see if they were fresh, and then the cook would make us omelettes, which made a change from the goat-steaks. The boat always left at first light.

Later, we arrived in Stanleyville – a place-name my father also accepted – which is called Kisangani today. It was very hot there and I saw fishermen using nets on stilts to catch fish. There were no timetables, as such, so we just had to wait for trains and boats until they came. In Kindu we took an old train to Kongolo from where we took another boat along the Lualaba River. My father was waiting for us at Munongo. He was limping along with a crutch made of a tree branch and was suffering from dysentery. His life was saved by his black corporal who looked after him.

Some of the black people where we lived had never seen a white person before. There were very few white people there. There was a Polish Jew called Ptaschek, and an English Jew called Juste. I met them again many years later in Elisabethville (Lubumbashi) when I was about 20. They were very nice – they slept in a tent so that we could live in their house. They ran a shop that sold every-thing, including tinned peas, carrots, corned beef and sardines. We had hundreds of tins of sardines because they made a change from eating eggs and chicken.

In the bedroom I saw my first snake – I had never seen one before. But Ptaschek smashed it with his gun. Because my father was living with his wife and child, he had the right to hire 50 porters. The locals fought to work as porters. People think they were slaves, but they were not; they were paid, fed and could take their wives along also if they wanted to. It took us six days to get to Mwema on foot. I had eight porters, working in teams, to carry me in a box. We needed all 50 porters because my mother's trunks were heavy and we had bought a lot of food in Munongo.

My father was very strict and would not let me be carried in the box if the porters were climbing a mountain. He was even accused of being too nice, but he was strict. He treated the locals like me, strictly, but he loved them all and they loved him. He treated them like children and that is what they accused us of, but how do you treat people who have never seen civilisation, eat sitting on the floor, and were not allowed to sit in front of their chief? The chief of a village would sit on a little chair but they had to sit on the floor at his feet. When a Belgian territorial agent or administrator came along in the beginning they might do the stupid thing and let the locals sit on a chair and so he would have no authority. My father would sit on a chair talking with the chief, but the locals did not seem to mind and did not really want to be equal in those days. They had not been to school, could not read and did not have any manners, except their customs, which were very nice really. In that part of the world, the tribes, including the Balubas and Lundas, were very jolly. They were always in a good mood, always joking, and never bore a grudge against you. Even if you punished them one day, they would come back smiling the next day. They were nice people.

My father had built a kind of mud-house for us in Mwema in the mountains, where it was very cold in the dry season. The house even had a chimney but we could not burn wood in it – we had to burn charcoal in a native pot for heating and cooking. My father's administrative duties took him on walking safaris for twenty-five or twenty-six days per month. The territory he was supposed to civilise was as big as a Belgian province, and he had to do that work alone. There was still one chief of a big village who would not accept the Belgians, and there my father was received with bows and arrows at the ready. He was very diplomatic, however, and managed to arrange things peaceably.

When my father went on his travels, my mother and I always went with him – and later my sister, too, who was born in Kongolo. The big canvas tent was erected and the cook would prepare meals. The tsetse fly was a problem because, while it carried sickness to animals, it had a sting like a bee. But the tsetse did not fly at night, so if we were going somewhere with tsetse flies we would leave very early to avoid being stung.

In those days, when the white man, or bwana, arrived in a village, he was always received by warriors who threw their spears just beside his feet. On no account were you supposed to run away because if you did, they would laugh at you for the whole time you stayed there. There was one Belgian who did that – he ran away and for years afterwards the locals laughed and sang 'Bwana kumvee, bwana kumvee. Anna kimbia, anna kimbia. Eh, hee, hee, bwana kumvee' (Bald Bwana, ran away. Ha, ha. Bald Bwana). The Bwana kumvee was an administrator who did not have a hair on his head – hence the name 'kumvee', which in Swahili is a rocky hill on which nothing grows. The locals had a sense of humour. It was amazing the things they called white people. So they laughed at Bwana kumvee for the three years he was there. My father even wrote a report saying that he should be transferred elsewhere. My father asked him why he had run away, and he said: 'Well, when I saw those spears coming, I had to run.' But my father told him: 'Yes, but you are supposed to stand there. The men are very good warriors and they know where they are throwing their spears.' When they threw the spears at you it wasn't a warning, it was a welcome. My father loved it, but some years later they did not do it anymore. Not all the tribes did that, but the ones we met did.

Most of the negroes did not have any real clothes. The missionaries gave them clothes, of course. The Catholic missionaries were richer than the Protestant ones, who were usually very poor. My father did not believe in anything but he helped the missionaries a lot because he thought they were a good thing. He would receive money to build one school in a village but, instead, he had three small schools built in three villages. But then he had to find teachers, and that is where the missionaries came in. The schools were for the natives. In the beginning, they did not want girls to go to school. The little girls did not even want to go to school because by the time they were five they wanted to help mummy in fields – it made them feel special. So, only boys went to school where they were taught to read, write and count. It was all done through Swahili – that is why I never went to school in the bush. It would have been no use to me then. I learned Swahili later, though. My father had studied it and I come from a multi-lingual family. My grandmother from Limburg spoke five languages, she even learned Swedish when she was about 60. An uncle of mine spoke eight languages, while my father was very good at French, Flemish and English, having lived in America. At home we would speak one language per day, and my father would insist that not one word of any other language was used that day.

Lots of people who came to the Congo later had tried to learn Swahili in Belgium, but got cross with their boys because they did not understand this or that. My father told them: 'But you don't speak Swahili properly.' For example, there was no Swahili for mouchoir (handkerchief) so that caused misunderstandings. I learned a local dialect, Kilubahemba, when I was three but it is completely different to Swahili, and only one tribe spoke it. It was like the vulgar language they speak in the port of Antwerp.

My father was involved in mapping routes for road-building projects. A few years ago, a man asked me angrily: 'Why did you have to take porters, why didn't you take a truck?' I replied: 'First you'd have to find a truck that can negotiate trees and elephants in the bush.' There were no roads and sometimes there was not even a track, so we had to use a compass to reach the next village.

My father had to collect taxes because once the locals got money for the things they sold, they had to pay a tax of one franc. He then used to put the money in little cases, which were not even locked. He would send two porters to the district town with the taxes to be handed over to the 'chef du territoire' or district supervisor in Munongo, because the nearest administrator was about five or six days' journey away in Manono. We called the black porters 'les coureurs' because they were very good runners and you can see who wins the international athletic events nowadays. They operated a kind of telegram service.

Every evening at 5 o'clock the Europeans would drink what the English call 'sundowners'. My father didn't drink any alcohol and we did not have any, apart from a bottle of whiskey in a Red Cross box. There were bandages and various things in the box, including the whiskey, which was labelled 'medicinal comfort'.

On one occasion, my mother went to the local Belgian doctor as she thought she was pregnant. The doctor examined her and said 'Oh no, you're not pregnant.' He was not a very good doctor, however, because as the months went on it was obvious that my mother was pregnant. Another time, my father was called by the army, the Force Publique, to a military school for black soldiers. All the officers were white. One of the black corporals had died and in those days the blacks did not want to be locked up in a coffin to be buried because they thought their spirit would never get out. They wanted to be buried in the ground, so the coffins for blacks had a removable bottom, which could be pulled out during the burial ceremony so that the corpse fell into the grave. In this case, however, when the 'corpse' hit the ground, the black corporal sat up and said 'Jambo Bwana' (Hello Boss). The same Belgian doctor, who had signed the corporal's death certificate a short time beforehand, begged my father to cancel the certificate. But my father said: 'I don't have a certificate for someone who comes back from the dead.' He then sent a written complaint to the governor and the doctor was kicked out. The Belgian medic was later replaced by a nice little Italian doctor. My mother went to see him and he said: 'My God, you're five months pregnant.'

It was a fascinating life in the Congo and I was a very happy child up to nine years of age when, in 1930, my parents sent me back to Belgium to attend primary school. The Belgians were very stupid in those days; they had no psychology or anything. They didn't think. I had been brought up like a little bush girl, bathing naked in the Lualaba river – which was full of crocodiles – with all the little native boys. It was all so natural, but my father did not earn enough to send me to a boarding school for white children in Lubumbashi, and it would have been no use to send me to a native school to learn to write Swahili because I would not have had any lessons in history or geography. People in Belgium thought we were millionaires in the Congo, but the people who worked for the government in Africa were very badly paid. So I was sent to school in Belgium from the age of 9 to 17.

Can you imagine a little girl who has lived in liberty with animals and things like that, being locked up with nuns – les Filles de la Croix? It was terrible. I was sent to church twice a day, and cried every night. Then I started to wet the bed so the nuns made me walk in front of all the other girls, carrying the wet sheet to be washed. My father finally agreed to take me out of the convent when my grandmother wrote to tell him that I was obliged to wear underwear when washing. Then I was transferred to a school in Antwerp – les Soeurs de la Charité, who were a little bit more modern – but I was kicked out because I knew that babies did not come from a white boat in the port of Antwerp. That's what all the other 12-year-old girls in the convent believed.

The nuns wrote to my grandfather to tell him that I would not be coming back to the school in September. That was because I had asked them how it was possible that Adam and Eve were the first people on earth, and I wanted to know who their children had married in order to have more children. The nuns said, 'Oh my God', and forced me to kneel in church with my arms outstretched for hours. On another occasion, I asked them how big Noah's Ark had been, because I had lived among animals and knew how big they were and how much they ate. I told the nun, 'The Ark must have been very big because two elephants eat a lot, and the lions must have antelopes to eat, so where did Noah put them all?' So, I was punished again because you weren't supposed to ask questions like that.

Then I was put in a non-denominational secular school, le Collège Marie-José, which still exists in Antwerp. There we had sex education, but it was very scientific. Whereas in the convent I had done very badly, I now started to do well, learning modern languages as well as classical Greek and Latin.

I returned to the Congo in 1938, while my sister Loretta stayed behind in Belgium. Before going back to Africa, I studied at the natural history museum in Brussels, learning to preserve and stuff small birds, fish, snakes and other small mammals, as well as insects like butterflies. My father was a friend of the curator who wanted me to send specimens back from Africa, which I did later on. In the museum, I even helped to repair a stuffed elephant, which had holes in one leg.

We lived in Kiambi where my father had worked since 1930. It was a small post but bigger than Mwema where I had been as a small child. There was an Italian merchant and a couple of missionaries. Kiambi was a very interesting area for tropical bird life. I had a teenage native boy to help me. He was good at killing birds, but would not touch snakes. I had to catch them – mostly vipers, which are not a problem because they are very slow and easy to catch. The museum in Brussels had forgotten to give me a special fork to catch snakes, so I had to catch them with my bare hands. I had to wear glasses when chasing snakes because they spit venom at your eyes. I also had to wear an apron otherwise the venom would corrode my clothes. The black cobra was very dangerous. You had to hit them in the middle with a stick so they could not advance any more and would rise up. I was terrified the first few times, but I learned to tease the snake to make it spit, and then I could put my hand around its neck. The black boy was standing about two metres behind me with a jar to put the

snake in. The following day we had to clean out the snake's insides before putting the skin in the jar of formaldehyde. We had to be careful not to damage the
head or tail, just cutting open the body with a scalpel. We found funny things
inside snakes, including half-digested lizards and birds. Inside crocodiles you
could find rings and pearls. My specimens were sent back to the museum in
Brussels where, I presume, they are still on display.

My father was strict with the blacks. Nowadays they say, 'You put chains on
the prisoners', and that is true but the chains were only used for dangerous people, such as murderers. In Kiambi, there was a jail but the prisoners loved it. They
had a good meal every day, although they had to work. I always said that you'd
need three houseboys to replace a Belgian maid, because they sing while they are
cutting the grass, and so on. There was a tribe in Kiambi called the Bambotes.
They were small but not as small as the pygmies. They were hunters and did not
have villages, so my father couldn't manage to see them so they didn't pay any
taxes and they didn't have any money anyway. They should have presented
themselves to him, however, for the census. One day, we approached a village
and two little children were crying. My father asked them why they were crying and they said, 'We brought meat yesterday to the village and they should
have paid us but they didn't.' My father went to the village chief and said: 'You
give those children the money you owe them because they brought the meat.'
So the children were paid and the next day all the Bambotes were there in front
of my father's office. They said, 'Bwana, we cannot pay taxes but we will go
to jail if we have to.' My father replied, 'You should really do fifteen days but
I do not have any room in the jail for you.' So the Bambote chief replied, 'That
doesn't matter. We will build a little village outside the jail and will stay there
for fifteen days.' They were not very good at digging or road works, so my
father sent them out hunting for meat. They made little huts and stayed there for
fifteen days. From then on, they did that every year but when my father was
replaced, the Bambotes were never seen again.

When local villagers served time in jail they were considered as sort of
saints – they were very proud of it. 'Oh, you haven't been to jail?' they'd ask
each other. It was like going to boarding school for us – to Eton or some place
like that. You could understand it because if the men were in jail they didn't
have to fish or go hunting for meat, and the wives didn't have to cook for them.

There was trouble with one sect from Cape Town – a bit like the Jehovah's
Witnesses – who turned around the biblical teaching so that Adam and Eve
were black, and Noah was black. They said we whites were the bad ones and
they were the good ones, so they had to kill the whites. One of those chiefs in
the Congo came from Kiambi and my father was ordered to arrest him because
such sects were not allowed in those days, probably because of the Catholic
priests who dominated everything. My father said that we could not do without
the Catholic clergy because they trained the student teachers. There were
Salesians and White Fathers who were the aristocracy of the missionaries. The
blacks learned Christianity from those catechists. One of my father's friends out

there was a posh English missionary called Crawford who gave me his book for my fourth birthday, even though I could not read. I still have the book, called 'Back to the Long Grass', on which he wrote 'My dear little Yetta'.

We knew one White Father who had been in the Congo since 1887. In 1940 he was a jolly old man with a big white beard. My father asked him if he thought he had converted many Congolese in over half a century of missionary work. The priest replied, 'If I come to the door of St Peter, I will be very happy if I have converted one.' My father said, 'But you've worked all these years,' and the priest responded, 'I know, but sometimes they are Catholic and sometimes they are Protestant. It depends on what little present that mission gives them this week.'

The locals were made to be good Christians and they understood it. They didn't have any religion really – they believed their ancestors came back as zebras or birds. There is one type of little black bird with a white tail that goes up and down – we do not have them in Europe – which is allowed to do anything because it is considered to be a person's ancestor. Nobody can touch those birds, so they can go into any hut and eat what they like. The locals had no religion but they could understand God.

When my father left Kiambi, he was known by the locals as Bwana Tchui, which means leopard boss. The locals acknowledged that since my father arrived, they had developed big fields of maize, manioc, peanuts, cotton and other products. He had organised cotton production in the region and the villagers were happy because they could sell their cotton every week as the truck came round to collect it.

In the early 1940s, there were 2,000 American troops in the city of Elisabethville because of the war. In one way it was a blessing because they modernised the airport and the roads. They behaved very well.

I don't know if I remember any change in the mood of the local blacks leading up to independence in 1960, because they were a very cheerful kind of people. When my mother came from the bush in Kabolo to live in Elisabethville, she brought her two boys with her, and they were happy to come. They had been years with my parents.

In 1946, I married Willy Cardon who was chief engineer in the Brasserie du Katanga in Elisabethville, the Belgian-owned brewery that made Simba beer. It had been built in the early 1920s. Whenever I had to phone my uncle in Belgium, I went to the post office with two bottles of beer because then I was put through faster. I worked for the South African consulate there for six years, and on several occasions I acted as vice-consul. My husband had two houseboys, one of whom was called Alexandre – he was a very good cook. Most Europeans had a boy to clean the house, make the beds and do the ironing. In earlier times, the cook would have a helper to peel potatoes and clean pots. I had another boy called Louis who wrote to me after we left the Congo; I still have his letters.

The independence movement started in the 1950s but the trouble was that in

Belgium they never asked the colonials what they thought about it. All the colonials thought the Congo had to become independent, but not so fast because central Africa was the last place to become civilised. They only had eighty years [of colonisation] behind them, but we needed 2,000 to be where we are. Some villages were so far away that the locals had never seen a car or a train. The black politicians hoped that the discussions in Brussels [in January 1960, concerning independence] would last – you know, palaver, palaver, palaver – about five or six years. Some of them told me that. But it wasn't like that. The Belgians said: 'You want independence? Okay, you'll get it,' but they weren't ready for it. We knew. I had known Tshombe personally, since he was a teenager.

They all hated Lumumba. I heard one of his speeches in Elisabethville. My two boys, Constantin and Louis, told me, 'We don't like that Lumumba.' So I said, 'You'd better go and listen to his speech.' Off they went, but I went too and stayed at the back of the hall. The things he said were horrible, and he was drugged. People don't believe me, but I have seen it. They smoked chanvre [hemp]. It was accepted as medicinal for old men who were sick. You would see their eyes becoming bloodshot. I listened to Lumumba's speech. He said, 'You've got to kill all the white men, but keep the white women. You can have all their houses and all their cars. When the men are dead you can marry the white women.' The whole speech was like that. I heard it myself. I wrote a letter about it to a local newspaper, Le Soir Illustré. They acknowledged receipt of my letter but didn't publish it. I went home after Lumumba's speech and pretended that I had not been there, so I asked Louis what the speech was like. He said, 'That man is crazy. He said that I have got to kill the Bwana and then I will get a white wife, but I don't want that.' You're not leaving us, are you? I don't want you to go.' I said, 'I don't know what is going to happen.'

It all went wrong. We knew the politicians but there was a funny mentality. People blamed us, the Belgians, because we didn't sit down at the table and eat with the blacks. In those days, of course not – they [the blacks] didn't even want to. In 1958, lots of hotels were open for blacks, so I said to Louis, 'Why don't you go and have a drink at the Palace restaurant?' He thought he was not allowed to go there but I said, 'Yes, you are now. You can go with Constantin and drink at the Palace.' When he came back, he told me, 'You can sit there politely with a glass in front of you but isn't it boring? I prefer going to a café in the cité [black township].' After Lumumba's speech, however, both my boys were very worried. They had seen my children born and had helped to bring them up. I used to go out at night and leave the children alone with the boys. I never ever thought anything would go wrong.

Every year a European circus came to Elisabethville, with zebras and clowns. I always sent my family with the boys. But the boys weren't impressed by a black dwarf in the show. They said, 'He is no good. A white man stands against a wall and the dwarf throws knives, but he misses him every time.'

The boys were like old servants in the olden days. They dressed in white to serve at the table and looked after the children. I loved them a lot, and I cried when we left the Congo [in 1967]. They cried also, and I got letters from them all the time. We used to go back every year – the last time was in 1971, the year my husband died. I never went back to the Congo after that, although I would have if I had been sure of receiving my pension out there.

We had been paying into a pension scheme for the two boys for years, but, as my husband guessed, they never got the pension when we left the Congo. So, before we left, my husband bought two small houses (with four rooms and an outside kitchen and toilet) for the boys. There was also a garden around so they could plant peanuts and other food crops.

At the time of independence, in 1960, it was dangerous for whites in some places. My Canadian aunt fled to Rhodesia at that time, and she took my son and daughter with her. My husband decided to stay, so I said I would not leave him. They started shooting and they killed the only Chinese man in Elisabethville. We were very fond of him; he ran a shop. He was stupid really because he took his car and went to the military camp, where they shot him. A few days later my husband came home and said, 'Look, you've got to go to Rhodesia with all the women of the brewery. None of them speaks English – you are the only one.' We left in the middle of the night for Rhodesia in a convoy of four cars, and arrived there at 4 o'clock in the morning. The British soldiers were waiting for us because they had expected it. For things like that the English are really wonderful; they had spare mattresses for us. At that early hour, the soldiers took us to a canteen and the English ladies were making sandwiches, tea and coffee. We couldn't believe it. The next day was a Sunday and the shops in Ndola were all opened for us. I had packed my bag in a hurry, so I needed to buy things. They let us stay in their houses, and my children stayed with a British army major. I stayed in another house with four other women and their children. Then I telegraphed the brewery's head office in Brussels for some money, which arrived two days later.

I rarely believe journalists because of the things they reported at that time. Every day in the canteen at Ndola we met more people who had fled Katanga. I asked one man how it was going in Elisabethville and he told me, 'Oh it's terrible. The blacks killed all the cats and babies by hammering them against the wall.' I thought that was impossible because no black, who is sober, would kill a baby. The man went on, 'On the road I met a man, and they had cut his arm off.' He claimed the one armed man was driving a car, but I accused him of exaggerating.

My father had been very friendly with Tshombe's father, because my father had been in charge of the black cité. Tshombe's father, who was a businessman, had been the first black man in Katanga to be issued with an identity card. Sometimes on Sundays, my father would say, 'I'm going to have coffee with Tshombe, do you want to come?' I was very surprised because my father sat at the table with Tshombe and we drank coffee and had a chat. Tshombe, the son, was about 17 when I met him. When he became president in 1960 he was a very nice man but the trouble with the blacks is that they are very much influenced by their family and tribe. So, even if Tshombe was honest and good, he had what they call 'l'entourage' [his inner circle] who influenced him. They told him, 'You mustn't give that money. Keep that money', and so on.

One day, when Tshombe was president of Katanga, my husband Willy met him and said, 'Can I invite you to dinner, President?' Tshombe replied, 'Yes, sure.' We invited a few people, including the boss of the brewery. I decided to tell the boys who was coming to dinner, and my boy Louis said, 'What? He is going to sit at the table with you? I'm not serving him and I'm not cooking for him.' Constantin said he would not serve Tshombe either. The problem was that Tshombe was from another tribe, the Lunda, and there was also a class difference. My grandparents, for example, would not have invited a miner from Charleroi to dinner, and things were still at that stage in the Congo. Luckily, however, Tshombe phoned to say that he couldn't come to dinner. The boys said Tshombe was a good man but they couldn't understand that we would sit down at the table with him. That was going too far for them.

The Katangese independence movement was happening too fast. My father was a father figure to the blacks. From 1940 to 1948, he supervised the construction of the black township in Elisabethville. There had been awful huts there before. To the people who blame us for what happened, I say, 'What do you think? Wouldn't a black woman with four kids prefer to have a father who provides, than a government that doesn't provide anything?' They had hospitals, which we had built. Rhodesian doctors came to Kiambi and other places, to see the hospitals and dispensaries. There was a dispensary in every outpost. When there was no white doctor, there would be a black nurse – who was always a man. The blacks became very good doctors later – a bit like the Indian doctors in England.

Later on, UN soldiers were fighting against Tshombe's Katangese forces. The UN troops messed up my house. The Swedes shot through the windows and killed my neighbour. The Ethiopians killed a couple of friends – a man in his 60s and his 85-year-old mother. They killed their boy, also. The other boy managed to escape by hiding under the kitchen sink. The UN also bombarded us so we moved out of our house and went to the brewery boss's house. He had left for Belgium so the house was empty, and it was opposite the hospital. There was a big red cross on the hospital and we thought the Swedes wouldn't bombard us but they did. They went on bombarding us all the time, with planes, in the middle of the night. One night I counted 2,000 rockets, which hit the area.

During this time, I baked bread in my mother's old petrol stove, because there was no electricity. I'd give a loaf to a neighbouring family, and half a loaf to a couple nearby. During the day there was less shooting. The missionary priests lived next to the hospital and they had a lot of food, so my husband went to see them to ask for some food, and they gave us meat. We had no fridge so I cooked the meat every day and we used to eat a whole antelope in a week. Nowadays, food products have a sell-by date, but I never pay any attention to it. In the Congo, we never got food poisoning.

The UN Gurkhas were trained by the British army – they demolished things, but always under orders. They did not steal anything; they were very disciplined. My aunt was living outside Elisabethville and one day she heard shooting, so she went down on her knees and peeped out the window. Then a Gurkha popped up and smiled at her through the window. You couldn't have done that with the others, except the Irish who would have been drunk anyway. The UN troops lived in my mother's house and when they left we found that they had sold the bath, the doors and everything else.

The Italians worked for the Red Cross, they didn't fight. Italians don't like to fight anyway – the Belgians beat them in Abyssinia at the beginning of the war.

When independence was declared on 30 June 1960, we had to give up all our guns, but we did not really want to surrender them. I had a Luger from the 1914–18 war that belonged to my father, and I was very attached to it. I also had a Colt that my father bought in America. He took it with him on a trip to Mexico because in those days they were fighting all over Mexico. I had my own gun that I used to shoot snakes and birds. My husband also had a gun he used to hunt crocodiles, and we also had a .22 rifle.

Union Minière was the biggest mining company in Africa, with huge copper reserves. It was started by South Africans in the early 1900s. In those days Stanley tried to turn the Congo into a colony for the British, but they refused. Afterwards he worked for the Belgians. King Léopold II did a lot for Belgium, it's true, but otherwise he was a rather hateful man. His wife and daughter hated him, for example. He was punished because his only son died as a baby. His nephew, who replaced him on the throne, died when he was only 20 years old. His other nephew, Albert, became king although he had never been brought up to be a monarch. He was completely different and didn't even like his new role in the beginning. In those days, the South Africans came to run the mines. Union Minière had copper and uranium mines, as well as some diamonds. The tin mines were in Manono. We all presumed that Union Minière arranged everything so that Katanga would become independent. The Congo is an enormous country with many tribes. It would have been more logical to have several countries there but, of course, the riches were in Katanga, as well as in Kasai province where the diamonds were. Between Kasai and Katanga, there was practically every mineral, including cobalt.

We were disappointed when the secession of Katanga did not work. Firstly, we liked Tshombe and, secondly, we thought that some big country was at the back of this. When America decided that it could not let the secession happen, there was nothing Belgium could do because it is very small. At the time, Belgian youngsters, aged 18 or 19, joined the black Force Publique to fight against UN forces for the independence of Katanga. The UN thought it was going to be very easy; they did not know that the Katangese army would fight back like they did.

The Swedish had not had a war for so long, I think they were just afraid. At the beginning of hostilities, we slept in the bathroom because it was the only room with a cement ceiling. For three nights my husband, myself and the two kids slept like that, but on the fourth night I decided I was going to sleep in my bed. The next morning I awoke to find all the windows had been smashed in the fighting, but it had not even woken me up because I was so sleepy. All my curtains and armchairs were ruined by flying glass. The Swedish used troop carriers at night and they shot at anything that moved. Quite a lot of blacks were killed, but the blacks killed a lot of Swedes too. We did not see much of the Irish soldiers; they were mainly around the airport. The Gurkhas were positioned on the flat roof of the main post office, from where they could see the whole town centre.

Things calmed down afterwards. I was secure because I had three German shepherds in the garden, and blacks are very afraid of those big dogs. The guard dogs didn't let anyone in – white or black. Finally, we buried the Colt and Luger pistols in the garden in the middle of the night, and hid the rifles in the attic and garage. My husband was on good terms with the blacks and got on well with them. No one from the brewery was ever attacked, probably because they wanted the beer supplies to continue. No one ever came to check if we had guns in the house, and the boys never talked about it so no one found out.

Many Europeans left for Belgium and elsewhere and had let loose their dogs, which roamed the streets in packs attacking at night in search of food. They were very dangerous. Eventually, the mayor sent some soldiers to shoot them. It would have been more charitable of the owners to kill them before they left, rather than letting them loose without food.

During all that time, we were able to listen to Radio Collège, run by the Catholic College in Elisabethville. Just like during the Second World War, we were able to get news in English from Radio Mozambique. As a Portuguese colony, Mozambique wasn't in the war, so we always had the up-to-date news. Officially, as a Belgian colony, the Congo was at war in the 1940–45 period, and that is why we had so many American and South African soldiers there. Some Belgians went to South Africa to train as fighter pilots – some of them went to Egypt, and others to Italy where they were killed in the fighting.

The first atomic bomb was made using uranium from Katanga. The uranium from Katanga was used for medical purposes also, such as x-rays. One of the Belgian mining engineers had predicted that the Americans would make the bomb and was involved in shipping the uranium to America for that purpose. One can apportion blame but it was a choice between killing Japanese or killing Americans. Without the first two atomic bombs, the war would have gone on and on, and more Americans would have been killed. That's war.

We left the Congo in 1967 because my husband was promoted to be a director in the Brussels head office, which was very good for his bank account. After leaving the Congo, we went back every year. We saw everything deteriorating because blacks moved in to a lot of houses that had been left empty. They had never lived in houses like that so twenty of them would go in at a time with a horde of children who wrote on the walls. They cut all the trees around the house to burn in the fire and the electricity broke down all the time. They would show films at the cinema, but would put in the last reel first, so nobody understood anything. Coastal areas of Africa were civilised a long time before Katanga was. The Portuguese had travelled through there in the 1500s and that is why a lot of the place names in the Bas-Congo are Portuguese.

I spoke Swahili, but in the French colonies the Europeans never bothered to learn Lingala – they thought all the negroes should speak French. I do not think that is right because they have their native languages. Swahili is made up of many different languages, including Arabic, Portuguese and English. For example, a bed-sheet in Swahili is called a sissy-betty.

As regards the stories about atrocities during the reign of Léopold II, he was not personally involved, of course. I agree that he was after money, certainly, and, as a man, he had so many mistresses. His father, Léopold I was a calmer person, and came under the British influence as he was married to a British woman – an aunt of Queen Victoria – who died in childbirth. He was remarried to a French princess, and had a son, Léopold, and two daughters by her. He used to visit England often and stayed with Queen Victoria who was his niece. He influenced Victoria very much. It is because of Léopold I that she met Prince Albert, who was German. Léopold II changed everything – he organised the construction of the Avenue Louise in Brussels, and the Royal Gardens at Laeken. He did a lot, including supervising the construction of beautiful buildings. He put all his money into the Congo, it was his property, not Belgium's. He sent men down there who treated the blacks as if they were animals. But do you think all the Belgians at the time were saints? In those days, little boys of ten were sent down the mines or cleaning chimneys, whereas in the Congo, we were not even allowed to hire a boy under 16 years of age. There are always bad people everywhere. Most of the intellectual Congolese are very moderate about this. I remember when I was about 18, some British doctors came from Kenya. They told my father, 'You know, this country is amazing. We can't understand how little Belgium has built so many hospitals and dispensaries.' In every big town there were two hospitals, one for the blacks and one for the whites.

Perhaps I'm just being philosophical, but I think we couldn't bring these people into the twentieth century in such a short time. I think back to that black corporal who saved my father when he had dysentery, you see how they were. Of course, they make them drink or take drugs to make them fight. But they are the same as all other human beings, and they even have qualities the whites don't have, such as the way they bring up their children. Everybody says the negroes are lazy, but they are not; they are slow because they have the time. They might say, 'Why worry? Why must this road be finished by tonight? It's silly.' They would get up with the sun and walk six or seven kilometres to work, and they'd go to bed at sunset.

In the 1950s most people were rather friendly with the blacks. Even after all those years in Africa, there are some things I still don't understand about them. We have another mentality; we are different. For instance, they are not allowed to say 'No' because it is impolite, so they may say 'Yes' when they mean 'No'. It took the whites a long time to grasp that you have to turn the meaning another way in order to understand it.

I don't feel 100 per cent Belgian after my years in Africa because there are things I can't understand. When I came back in 1967, I couldn't understand two ladies discussing the price of butter, because what did it matter if it was one franc more or less? All the Belgians who were in the Congo have become like a family – friends for life. The other day I attended a dinner of ex-colonials and saw an old man who had lived in Albertville. I asked him if he had known a Doctor Fonsney, and he laughed and said, 'Yes. Better to call him in the morning than in the evening.' Of course, there are fewer and fewer people at the annual dinner for ex-colonials. There are fewer tables every year. It's getting smaller and smaller.

15

Fighting For Our Lives
With 'Jadotville Jack'

Pat Dunleavy of the Organisation of National Ex-Servicemen and Women pays homage to the late Colonel Patrick J. Quinlan, who commanded Irish troops at the battle of Jadotville, Katanga, in September 1961.

I joined the army in October 1960 and did my initial recruitment training in Athlone Barracks. The first Irish battalion – the 32nd Battalion – went overseas in that period. I was posted back to Mullingar and in April 1961 we volunteered to go over with the 35th Battalion. I was with A-company in that battalion, who were mostly all from the Western Command, with the exception of one or two from the Eastern Command – they were signallers, armourers and artificers. We went out in June 1961 and Commandant Jack Quinlan was our commanding officer. Before we left the square in Athlone Barracks to parade through the town, he said that we were all going out together and would all come back together. He stuck by his word and we all came back, despite all the troubles we went through out there.

16. Pat Dunleavy (back row, first left) with Irish army comrades in Elisabethville, July 1961

At the beginning, we were based in Elisabethville before being sent to Jadotville. From there we were posted to Kolwezi and then back to Jadotville. We were welcomed to Jadotville by the population at the time. We were not fully aware of why we were being sent there as a platoon. Our commanding officer, Jack Quinlan, went to meet the mayor of Jadotville to inform him that we were there as a peacekeeping mission to keep the peace within the area, observe any problems that might arise and liaise with UN headquarters.

The 32nd and 33rd Battalions had encountered hostilities in the mineral-rich province of Katanga. Tshombe had been elected president of Katanga by Belgian officers, because the Belgians wanted to hold their stake in the mines there. They didn't want to share that wealth with the rest of the Congo. That is how the trouble started. Patrice Lumumba was the country's first prime minister and he wanted the Congo to have an equal share of Katanga's wealth. When we got there Katanga had already broken away from the Congo. There were various tribes there, including the main tribe of Balubas, who were fighting among themselves. The Katangese army was trying to quell those troubles and anyone in uniform became a prime target for the Balubas who lived in the bush. Hence the problem arose with the 33rd Battalion with the incident at Niemba. They were out on patrol from Niemba, to try to keep the dirt-track roads open, so these patrols could take place. Anyone in uniform was the enemy for the Balubas – the Balubakats could not understand who or what they were. On 8 November 1960, Lt Gleeson's patrol went to keep the road open and discovered that the makeshift bridge had been dismantled. They were a group of engineers who did that type of repair work. When they started work on the bridge they noticed a huge force of Balubas coming from all angles out of the bush to surround them. The big problem for us out there was the language barrier. We were not up to speed in Swahili, although we knew a few words such as 'Jambo', which means 'Hello, we greet you in peace'. French was also spoken, but not out in the bush.

Looking back, this was our first ever overseas involvement – it was a whole new concept for the army. We had never experienced anything like that before and, unfortunately, we learned by our mistakes. It is a long time ago.

The 34th Battalion came out and had a peaceful enough time, but the big upheaval took place in Elisabethville during our time with the 35th Battalion. Other UN contingents, including the Swedish and the Gurkhas, were in Elisabethville when trouble erupted with the Katangese army, of which Tshombe was the supreme commander. We were sent from Elisabethville to Kolwezi and Jadotville to keep the link open between those mining towns and the provincial capital. In normal conditions it would take about two hours to get from Elisabethville to Jadotville on a tarmacadamed road; it was not a dirt track one.

We were based in a farm and a factory in Elisabethville, and did a lot of patrols around the oil depots to make sure they were not blown up so there would be a reserve of oil supplies at all times.

The Jadotville incident came about as follows. We were not told why we were being sent there, it was just as a peacekeeping mission. Our commanding officer, Comdt Jack Quinlan, had presented himself to the lord mayor of Jadotville. He went there to see what the situation was like and was greeted by the whites fairly decently. When he came back he said he did not feel at ease so he made us dig trenches surrounding our area. We had to secure the area, which was about one and a half miles to the east of Jadotville town. We had a clear view of the mining area and could see the Katangese army paratroopers who had a base ten or fifteen minutes away from us. We noticed a very big build-up of troop movements in and around Jadotville so, naturally enough, our own O/C was in communications with HQ at Elisabethville to tell them about the situation. He was called in to Jadotville again and was not happy with the meeting he had there – they threatened him that if we did not move out of Jadotville and back to Elisabethville, hostilities would erupt. Quinlan came back and spoke to his commanders. Then we were ordered to reinforce our area by building new trenches, etc. We reported back to unit headquarters and were told to hold our ground. On the morning of 13 September 1961, as we were all going to Mass, the attack took place. We manned our trenches and fighting took place. We spent the following twelve days in trenches and the only time we got out was at nighttime. Ceasefires were called and Quinlan met the lord mayor of Jadotville again and asked him why these hostilities were taking place against us, because we were peacekeepers. The mayor was an African tribal chief but he was living in a town; he wasn't a culchie or a bushman. Trouble had also started in Elisabethville, so there were two conflicts going on at the same time. We were a company out on our own, with three platoons. We did not have sufficient ammunition, food or water to sustain us for a long period. The route from Elisabethville to Jadotville was cut off at the Lufira bridge. A relief company of Irish and Gurkha troops was sent from Elisabethville to relieve us but they couldn't take the bridge. On several occasions, they had to turn back. Two or three Gurkhas were killed in bombings there. There were jet attacks at all times by the Katangese paratroopers, which fired cannons and bombs into our area. We had no air support whatsoever, even though we were informed later on that the USA had been requested to supply Canberra bombers. The UN would not run with the plan because they did not want it to build up into a large scale attack, that would be reported all over the world. Eventually, however, it turned into a huge, worldwide media event. The request for the bombers was made by Dr Conor Cruise O'Brien, under the guidance of General Seán McKeown, who was the UN chief of staff out there at the time.

All our communications from Elisabethville to UN HQ in Léopoldville were done through the medium of Irish. Quinlan pleaded with them to send support because he could not hold out for much longer. We had been fighting constantly for fourteen days. It was like something we used to see in the cowboy and Indian films years ago, it was funny – once it got dark the fighting stopped. The Red Indians were afraid that if they were killed in action at night their souls would never go to the happy hunting ground. We thought these guys might have the same superstition that if they were shot at night they would never rest in peace.

We lasted for fourteen to fifteen days, before running out of ammunition, food and water. Without food and water you will not survive in the Congo. There were 130 men altogether in the company, against overwhelming Katangese forces. We were dug into trenches about a mile east of Jadotville, looking straight across dead ground to the mining area. We were attacked from the mining area by Belgian officers and the Katangese army, who fired at us with mortars. We returned fire but our mortars were superior to theirs. We inflicted casualties – thirteen Belgian officers died and there was a huge number of black casualties also. We sustained only minor injuries, no fatalities. There were bullet wounds to arms and legs, as well as shrapnel in the buttocks and legs from bombs. Some thirteen or fourteen of our men were injured in the fighting. I escaped injury.

Ceasefires took place to allow negotiations. At the end of the day, in Elisabethville, the UN took a huge amount of Katangese army prisoners. It was agreed between the Katangese army and the UN that we would surrender in Jadotville. The reason we surrendered was that no one could get to us with food and ammunition. We had no ammunition left. An agreement was signed that there would be an exchange of prisoners and that our weapons would be handed over. We rendered useless whatever weapons we had prior to the surrender. For example, the Gustav guns had a fixed bolt with a fixed striking pin, so our armour artificers filed down the striking pins. If you tried to fire them then, they could not penetrate the cap on the bullet to discharge the cartridge. They did the same with our FN rifles.

On the morning of the surrender and hand-over of weapons, the Belgian officers of the Katangese army said they would look after our wounded and bury the dead. When they were informed that we had no dead, they just would not believe us. They ordered us to show them where we had buried our dead. They were frustrated to discover that we had sustained no fatalities because we had inflicted terrible losses on them. That is what it was about, actually.

We were taken prisoner and brought to these camps along by the lake where the soldiers lived with their wives. It was very frightening. In Ireland and England when you want to get at someone, you boo them but over there it was totally different; they made war cries, threatening to cut off your testicles and shove them in your mouth, all that caper. It was a scary time. We were to be brought to Elisabethville for an exchange of prisoners. We went twice but the whole thing fell through, so we were brought back again. We were put into a disused hotel in Jadotville and were strip-searched. If anything was found on you that would be threatening to them, people got punched and kicked. We protested strongly that we had to be treated in accordance with the Geneva Convention, and respect was shown to us then. They understood what we were saying because they were controlled by Belgian officers. Some of the Belgians spoke English, and some of our officers spoke French. We were treated differently after that.

Without a shadow of a doubt our officers – and, in particular, our commanding officer, Comdt Quinlan, who afterwards was nicknamed 'Jadotville Jack' – were towers of strength. Quinlan showed great courage under fierce strain and pressure. Everyone who served with A-Company of the 35th Battalion would hold Quinlan in the highest esteem. He was our hero, he was great. Unfortunately, he was wrongly treated back home and we could do nothing about it. He was wrongly treated by his own superiors in Elisabethville and, afterwards, back in Ireland. He had a huge responsibility on his shoulders. He was responsible for the safety of his men. He had us out at all hours of the night because he would not be happy with the way this or that trench had been dug. He would say: 'It's not giving you proper cover here. You can't see out there.' He would make us dig more trenches. I'm telling you, digging out there with a little shovel and a pick, was like trying to dig through four feet of concrete because the ground was so hard.

On two occasions we were sent towards Elisabethville in a convoy of buses for the exchange of prisoners, but it never happened. Something went wrong. We were on the buses accompanied by armed Katangese personnel. The convoy was flanked by armoured jeeps. The negotiations broke down twice so we were brought back to the camps again. A decision was taken that if we were not released on the third attempt, one way or another we would make a break for it. We aimed to disarm the guards in the front and back of the bus and try to shoot the people in the land rover between us and the bus ahead. There would have been a lot killed but at the end of the day that was the decision that was taken. But luckily enough it did not come to that because the third attempted exchange was successful.

We were held for about four weeks in Jadotville during which time we got great food. All we usually had was rice, dog biscuits or whatever the cook could rustle up with what provisions he had. In fairness to them, the cooks were brilliant. We got good provisions in the prison camp in Jadotville. Some of us were in a prison camp in Kolwezi and we were treated better there than by the United Nations – unfortunately, I have to say that.

The negotiations for our release continued during the four weeks we were held. All we were allowed to do was walk around the compound to exercise. There was not much interference from the people who were guarding us. Local and international reporters came to get photographs and do interviews, but most of them were prevented from doing so. One or two interviews were done for the BBC World Service at the time, but that was it. We attended Mass every morning in our compound, celebrated by our chaplain, Fr Fagan. We had our own medical orderlies who checked that we were okay. If there was any need for medical attention our doctors were on hand at all times. The medical officer was Lt. Clune, who ended up later as director of the Medical Corps at St. Bricin's Hospital. We also had a Sgt. Harry Dickson. There were three or four corporals and two private medical orderlies as well.

The talks to secure our release went on between Comdt Quinlan and the mayor of Jadotville, and Tshombe's people in Katanga and the UN in Elisabethville. Eventually it was agreed that there would be a swap of prisoners and that hostilities would cease. On one occasion, Quinlan returned from Jadotville and told us: 'We are in for it. The lord mayor said that he was sending out an uncontrollable mob that would eat us up.' Quinlan's reply, delivered in his best Kerry accent, was: 'Send them out. We'll probably give them indigestion.' We took the mayor's threat as a joke. It was an attempt to get us to comply but we were not going to do that. We did not know why we had been sent there, although I am sure Quinlan was told at a later stage. Our job was to stay there until we were withdrawn.

We had a Swedish interpreter who was attached to the battalion, and we also had a guy called Mike Nolan. He was a former Irish mercenary and had been in the Congo long before the troubles ever started there. He was an old man at that time. He worked as an interpreter for the UN in Katanga.

Prior to these events, we were up at the crack of dawn to raid a barracks in Elisabethville. When we raided the barracks they were all in bed. We arrested the Belgian officers and took all their equipment. We also arrested President Tshombe at a checkpoint in the city, during the last week in August 1961. He was detained but later released, and that is when all the hostilities broke out. Later on, the Gurkhas stormed the post office in Elisabethville, where the Katangese army was based, and killed everyone in sight. They are fearsome fighters.

We were led to believe that our presence was to stop the breakaway by Katanga. The Katangese did not want the wealth to go to any other part of the Congo. The Belgians brought Tshombe to Belguim and got the biggest blonde lass in the place. He married her and brought her back to the Congo. He had more than one wife anyway. Tshombe was against the UN and was dictating this and that. He was against the rest of the Congo also. He was threatening violence against the UN at all times. If he had been held captive the hostilities may not have occurred, but we don't know.

We were guarding the Cruiser's [Dr Conor Cruise O'Brien] house in Elisabethville and Máire MacEntee was with him. I remember that the hottest time of the day is from 12 noon to 2 p.m. and we were acting as his personal guard but we could not use the swimming pool anytime during the day. We had to be out during that particular period guarding the house, while everyone else took a siesta. Cruise O'Brien was not flavour of the month with the Irish UN people. A lot of people blamed him for the whole situation that started there. We believe it to this day that he was the cause of all the problems in the Congo. I think he was relieved of his post later on. He had a private villa with its own grounds and swimming pool, which we were guarding. He had his own UN driver, a fellow called Trooper Wall from Clonmel. Wherever he went he had an armed UN escort. I never met him personally. He was there when we were on duty but he never spoke to us and we never spoke to him – it wasn't our job. We were withdrawn from those duties after a fortnight with him.

A-Company of the 35th Battalion got every kind of dirty detail that was handed out. We went on a five-day mission along dirt-track roads to Angola to pick up a Congolese minister and his wife and family. He was afraid to travel back into the Congo on his own so we brought him back to Elisabethville under UN protection. It took ten days altogether to do that. It was a great boost for us because it was a kind of safari. We slept in missions at night and washed in rivers. It was a great break, a sort of getaway.

A Norwegian helicopter pilot flew in from Elisabethville with provisions for us. We put up a blanket of fire, with sheets spread out to give him a landing base. But all he had brought with him were two containers of water and a few thousand cigarettes. There was no ammunition or anything else. We were surrounded by Katangese forces and that was all he brought! My God, it was a laugh. Even now, I'm not too happy with what the commanders did. Jesus, first of all, being sent out to tropical Africa in bullswool uniforms. No one in their right mind would have done that. It just goes to show you how ill equipped and inexperienced we were. They answered the call by the UN but didn't know what they were letting themselves in for. They were costly mistakes that should never have happened. I was 18 when I went out there and I'm 62 now. I was sent out after doing only six months' duty in Mullingar. Without a doubt, we were too young to realise the dangers we were involved in. I suppose the people at the top should have realised – as we call them, the people with the custard on their caps.

A-Company was a great company to be with. We had great officers, and our hero was Jadotville Jack Quinlan. I would never let anyone say a bad word about him. On the barrack square in Athlone, he told each and every one of us 'We're going out together and we're coming back together.' Lord rest him. He was badly treated back home. It was the white feather; they blamed him for surrendering even though he had no other option. The people who say those things about him had no experience of what went on out there. When we came back, they said: 'Oh, you surrendered. You laid down your arms' and all that. In my opinion, they were thick, ignorant gobshites who were never there. They didn't know what went on. They didn't know the true facts. They didn't know the consequences that would have occurred if we had persisted to fight with what? – nothing but our knuckles. It was unbelievable luck that we had no fatalities.

When we were released, on 25 October 1961, we were brought back to the farm where we had been based originally. We were sleeping in a cow shed with camp beds. You'd be eaten alive by mosquitoes. We got this stuff to rub on the exposed parts of the body at night but every bit of your body would be exposed at night because the heat was unnatural. You could not sleep with anything on because it was so warm. You'd be in a lather of sweat.

We never saw the other people that were exchanged for us, because the hand-overs happened at different locations. We were brought to the old airport in Elisabethville to be exchanged but I'm not sure where the others were brought. Lt Col Hugh McNamee, the battalion O/C, welcomed us back to the farm. He had taken over from the previous O/C of the battalion who had to be relieved of his post due to a drink problem. We had been there about a month when that happened.

Comdt Quinlan came home and served as O/C of the 2nd infantry battalion for a good few years. After a long, long time he reached the rank of colonel and was posted to Athlone. He retired and went to live back in Athlone. He was a Kerryman but had done most of his army service with the Western Command. He died about five years after retiring and is buried in Athlone. The men of A-Company are still very loyal to his memory.

The 36th Battalion came out to relieve us in December 1961. We arrived back in Dublin airport on 22 December and were brought to Clancy Barracks where we handed in our weapons. We got ten pounds for which we had to sign. That was a lot of money then. We also got a meal. We were sent to Collins Barracks but were not allowed outside the gates and could not talk to anyone. We were put on the train the next morning at Westland Row station, and arrived in Mullingar around 10.30 in the morning when we whipped straight up to the local barracks. We did not see our families until later that day. We had ten pounds for the whole Christmas period. We returned to normal army duties after that. I was with the transport unit. I did not go back to the Congo but served in Cyprus, which was totally different. I retired from the army in 1981.

16

Remembering Jadotville

Lars Fröberg served with the Swedish Army's 12th and 14th Battalions in the Congo and, for a time, was seconded to the Irish Army as an interpreter. Fluent in French, Lieutenant Fröberg fought alongside Comdt Patrick 'Jadotville Jack' Quinlan against Katangese forces at Jadotville in 1961. He acted as an interpreter for the 35th Battalion during the tense negotiations when 155 Irish troops were held captive at Jadotville and Kolwezi for almost six weeks. This is his story.

It all started on Sunday morning, 10 September 1961. I had been working hard and was very sleepy. I was planning to take it easy and relax, when the stillness was broken by a crunching sound outside the interpreters' villa on the Avenue des Savonniers in Elisabethville. They might have been the steps of fate I heard, although I didn't know it then. Somebody was obviously coming to pay us a visit. It was a Swedish corporal sent by order of the Swedish duty officer, Captain Carlstrand, with the following message: 'A French-speaking interpreter should report to the Irish battalion HQ as soon as possible.' I asked the corporal if he knew anything about the mission and he said we were going to Jadotville, an important mining town 120 km from Elisabethville. He added that I'd better bring a blanket 'as we just might have to spend the night there'.

Since I was the only interpreter available, I got dressed and went to the Irish battalion headquarters. After a good breakfast there, I left for Jadotville with the Irish unit, which was a reinforced platoon. I sat in the front of the jeep between the commanding officer of the force, Captain Donnelly, and the driver. The sun was shining and I was beginning to feel quite pleased about this 'little trip', although I had no way of knowing that I would not return for another forty-five days.

Following us in the convoy were two lorries transporting the soldiers, and the doctor's private car. I was fortunate to be at the head of the convoy as thick, red dust swirled up in clouds behind us. After a couple of hours on the road – as I realize now, in retrospect – the first warning came. I saw something moving right in front of us, so I shouted 'Stop!' 'What's the matter?,' asked Captain Donnelly. 'Have a look yourself,' I said and pointed. What we could see ahead was the Lufira bridge. On and around it were swarms of black soldiers – gendarmes. All our vehicles had now stopped. On our side of the bridge there were trucks parked in zigzag fashion and in front of them oil drums, filled with gravel and stones,

17. Swedish Army interpreter Lars Fröberg (right) with Irish troops in Elisabethville after their release from captivity on 25 October 1961

had been placed as a barrier. Later, we also saw white officers – Belgian and/or French.

Captain Donnelly seemed a little dubious about the situation, and I understood how he felt because this was a real barricade. The captain looked puzzled... 'What shall we do?,' he asked me. 'Just give me orders,' I said. Then he asked me, 'What would you do?' I suggested that we should let the soldiers stretch their legs and have a smoke break. In that way they would be clearly visible to our adversaries on the bridge. 'But no hint of aggression and absolutely no shooting,' I added.

'And you?,' Captain Donnelly asked. 'I'll go and talk to them,' I replied. 'But what will you say?,' he asked. 'I have no idea,' I answered, 'but we just can't stay here doing nothing.' Then I checked that my pistol could slide easily in and out of its holster, and headed off. As I strolled along the road towards the bridge, two of the Africans (Katangese gendarmes) left the roadblock and one of them lay down in a ditch taking aim at me with his automatic carbine. Meanwhile, the other one came towards me shouting and waving his weapon. I continued slowly, acting as if I didn't understand what was going on. Why all this excitement, and what was the man howling? Then I heard what he was saying: 'Arrête!' (Stop). I looked around me, trying to appear really silly, and I think I succeeded. Still I carried on. The black gendarme had come much nearer now, so I said, 'Parlez-vous français?' (Do you speak French?). Then he looked quizzically at me and replied 'Oui' (yes), while shaking his head from side to side, the way we do when we mean 'No'.

The man was clearly nervous and suspicious, so I asked him what his rank was. 'Caporal', he replied. I introduced myself as Major Fröberg of the UN contingent, even though I was just a second lieutenant! I told him we were on our way to an Irish company near Jadotville with medical supplies and that we had a doctor with us. I told him I could see that they, the Katangese, were on manoeuvres, adding 'Good. You should always be ready for action.' I told him we wanted to cross the bridge. Could he possibly help us to make it across? Then another very aggressive gendarme came up to us making vehement protests against my request for safe passage, so I asked the corporal which one of them had the highest rank. The corporal said he was in charge and that the other soldier was only a private. So I told the corporal: 'Order him [the private] to leave at once, before something unpleasant happens. Listen, I know your president [Tshombe] and he will certainly be informed of whatever takes place here. We have his permission to move wherever we want in this country. Katanga is not at war with us [the UN] and yet you are trying to stop us by force. Look, relax and have a cigarette.' He accepted one and after some more talking, I handed him the whole packet of cigarettes.

I went back to the Irish platoon and told Captain Donnelly what we had discussed. After a while we drove on in the jeep towards the Lufira bridge. In short, after some thrilling moments, they let us pass over the bridge. We were glad we had managed to get through the barriers; three men in a jeep were perhaps not considered too dangerous. No more of our troops were allowed to pass, apart from a truck driver and the doctor. We had been driving for about 30 minutes when my blue UN peaked cap suddenly blew off. The jeep stopped and Captain Donnelly sent the driver to get it for me. When we continued on our journey something else happened, which may have been a second warning. This time a gendarme armed with a rifle emerged from the bush, looking menacing. 'Well, what now?,' asked Captain Donnelly. 'Have a guess,' I replied. 'You'll go and talk to him,' said the captain. 'Yes,' I said. I went up to the black sentry, moving in a rather measured way, as I didn't want him to get more upset than he already was. He was very alert but I managed to get very near him, doing what I could to look friendly and understanding. One of the reasons for my moving up close was that he had a rifle and I only had a pistol and, if the worst came to the worst, I wouldn't like to miss the target.

I tried to explain that we had been let through at the Lufira bridge by the Katangese gendarmerie and that some of them would certainly be coming along any minute now. But he wasn't convinced. He took a step backwards and I took a step forwards, and so on, in what I call a 'danse macabre'. The man was looking desperate now and the situation was desperate, at least for one of us. Which one would turn out to be quickest on the draw? Just then we heard the whirring of a jeep and a truckload of black soldiers, the tension between us eased and the problem was solved. But what would have happened if my blue cap had not blown off earlier?

After this little *intermezzo*, we drove on towards the Jadotville area and, as there were no more obstacles, we arrived at the Irish camp in the afternoon. I have memories in abundance of my time with A-Company of the 35th Battalion in Jadotville and Kolwezi. Here is a snapshot: on Sunday evening, 10 September, it was getting dark when Comdt Quinlan and I heard some kind of howling in the distance. It seemed like human voices. At first I couldn't grasp the words but I soon found out by straining my ears to listen. It was a kind of chorus and the message was anything but encouraging. They were screaming over and over again in French: 'à bas l'ONU!' (down with the UN). Professional agitators must have been behind this – perhaps it was the third warning. I think the noise came from miners in the neighbourhood. Comdt Quinlan and I were not the only creatures to be disturbed by the screaming: dogs began to bark and – believe it or not – cocks began to crow at dusk. Then, after a period of silence, another sound emerged: the loud croaking of frogs or toads went on for several minutes… the atmosphere felt ominous.

On Monday, 11 September, Comdt Quinlan and I went into Jadotville to see the mayor who was called Amisi. There were two or three local mayors. He was a very nice African who appeared to be sensible and cultivated. Some gendarmerie officers were also present, including Major Makito, who was later promoted to the rank of colonel. He was one of President Tshombe's confidants and was also a sorcerer, but that's Africa for you. Another more intractable officer, named Tschipola, was head of the local gendarmerie (militia).

We were on pretty good terms with those in power and Comdt Quinlan and I went to a restaurant for a drink with Major Makito to show people that we could associate with them like civilized people. I thought this might ease the tension, especially among the white inhabitants of the city but, on second thoughts, I may have been wrong about that. The whites in the restaurant didn't look happy when we came in, to say the least. Some were very upset. I was relieved when we left Jadotville. We had to push our way through crowds of soldiers and it wasn't very pleasant as we were unarmed, or supposed to be. On no condition were we allowed to carry arms on our visit to Jadotville. Nevertheless, that Monday I had brought my little black Beretta. I thought better of it the following day, however, because if I had been searched and the Katangese had found a loaded pistol in my pocket I would not have been able to write these lines, or any others. There was no rioting, but we could easily see what the black soldiers would like to have done with us.

Back at the Irish camp that evening, some young Irish officers and I whiled away the time on the upper floor of an old petrol station, recounting ghost stories – each one more ghastly than the rest. That night I slept soundly on a camp bed, quite happy the day was over.

The following morning we had a rude awakening when a terrible crash brought us all back to reality. Bullets splintered the windows and whistled above our heads. The moment I heard the shooting, Lt Kevin Knightly bellowed from his camp bed: 'This means war!' I hope I'll never hear those words again. We rolled out of our beds, and remained horizontal. It was exactly 7.40 a.m. on

13th September and the war had begun. All the agreements and promises we had reached with the mayor of Jadotville had been broken. We hurried to our respective tasks, including Lt Knightly to his armoured cars, each with a Vickers machinegun. The attacking enemy would soon realize what that meant.

I rushed to the little villa on the other side of the Elisabethville-Jadotville road, where I had my interpreter's office. That was the only place with a telephone where the enemy could get in touch with us – apart from the real battlefield with trenches. But we couldn't ring them up, so I didn't like that telephone.

Then the shelling began. I was sitting at a window looking out, when the first shell landed. I felt quite superior. It landed very far from us, 60–100 metres away. I remember saying to the Irish sergeant who was sitting on the floor (clever of him), 'They shouldn't use modern weapons, but rather stick to the weapons they can use, like spears and poisoned arrows.' Then the second shell landed, but there was no explosion at all... it must have been a dud. Then we heard the crunching sound when the third one was fired; it would take about 22 or 23 seconds before hitting its target, but nothing was heard. I half turned to the sergeant and said, 'That was certainly not meant for...' – I was going to say 'us', but I was interrupted by a dreadful detonation just outside the window where I was sitting. The shell fell so near that its splinters smashed the window and I felt something sweeping past me, and my face was covered in dust. I groped through my hair to feel if there was any blood but there wasn't – just some gravel and dust.

Slowly I turned to the sergeant and asked him if he was all right. He didn't answer but just crouched down on the floor. He wasn't wounded but at first he couldn't understand how I could speak because he thought I was dead. The mortar fragments had missed my head by an inch or so and had crushed a wooden ceiling lamp in the shape of a ship's steering wheel. The sergeant later told me he had lost his voice for about thirty minutes. None of us suffered a scratch. What saved us was a concrete slab near the outer wall where the shell had landed in a dead angle about one metre from me. A narrow escape or what?

The next incident occurred just after this when I was summoned to a council of war at the old petrol station. It had to be important so I left for the place. I ran across the road – only an idiot would have walked with all that shooting going on – after shouting to the Irish post that I was coming, to prevent any mistaken identity. I lay down near two Irish corporals and had a few words with them, trying to cheer them up a little. Then I went on and was walking across a courtyard to the petrol station where the council of war was to be held. I couldn't see anyone but I heard the men in the petrol station shouting at me, 'Get down, get down'. Then I saw them gesticulating violently at the top of the steps. They seemed very excited and wanted me to take cover, but there was no shelter, just a stretch of bare ground. Seeing no danger, I proceeded towards the building and was almost there when I heard Comdt Quinlan's voice shouting: 'Get down. That's an or—' I only heard the first syllable of the word 'order' but if I hadn't acted then, I would have been shot dead. It was that close. I threw myself forward and before my hands reached the steps, two or three bullets hit the wall just where my head had been fractions of a second beforehand. Comdt Quinlan, Fr Fagan and some others saw it all from the top of the steps. There was a balustrade, which protected me as I swiftly crawled up. Talk about a close shave – that sniper almost got me!

Back at the interpreter's villa, another sort of war, a psychological one, was being waged over the telephone. The calls I received were a strain on my ears – different voices used various methods, but all of them were trying to bring us to our knees, to get us to give in. The most disgusting threats in coarse language were followed by the finest of promises. Sometimes we were told that natives in the surrounding villages would come and massacre us before tearing us to pieces to eat us. They said the Swedes and the Irish in Elisabethville had surrendered, and that the Katangese were just putting the finishing touches to slaughtering the Indians. On other occasions, however, we were guaranteed safe passage to the Rhodesian border and would only have to deal with whites. One caller claimed to represent the Red Cross, lamented the fact that we were so stubborn, and pretended that all he wanted was to save our lives. All we had to do, according to him, was surrender! – a generous offer? Then we were threatened with bombing by the Katangese air force's Fouga Magister jet. If we didn't listen to reason, the whole Katangese army would move against us, and so on.

This was a psychological 'war within a war', and these threats provide an idea of the prevailing situation in which we found ourselves. Given the circumstances, I tried to keep people's spirits up, but now and then conversation was cut short by violent outbursts of gunfire. It became harder to maintain our delaying tactics. All the time, Cmdt Quinlan and I were hoping that the promised reinforcements from Elisabethville would arrive in time, before we were completely exhausted by hunger, thirst and fatigue – the new enemies we now had to struggle against. No marksmanship can help you against hunger and thirst. Sometimes the soldiers sat sleeping over their weapons but the tension, during the constant attacks on us by Katangese forces, took its toll.

The nervous strain on our soldiers was horrible, especially after the Katangese Fouga bomber had been sent to break our morale. It was horrid to hear its whining sound coming closer and turning into an ear-splitting howl as it dived on us with its machine guns firing. We swiftly got accustomed to the jet's machine-guns, however, and I don't think anyone was particularly afraid of them. But when the Fouga dropped its first bomb, it was the beginning of a nightmare, which got worse and worse. Of course, we tried to shoot the plane down when it attacked us, and hoped for the best.

Most of those who led the fight against us in Jadotville were white mercenaries and there were plenty of them, even in Kolwezi. The mercenaries had been expelled from Katanga by the UN but it was not long before they returned to Katanga dressed in civilian clothing. They were, however, recognised by many UN soldiers.

On the second day of the battle, an Irish sergeant, named Ray, suddenly turned up in my villa with two white civilian prisoners. They had been caught in a roadblock we had set up on the road from Elisabethville and Jadotville. Both of them emphatically denied having anything to do with the Katangese gendarmerie, but their identity papers said otherwise. In addition, I recognized one of them as being connected to the gendarmerie. I had met him some time earlier on a warm, sunny Sunday at a big swimming pool in Elisabethville. Some gendarmes were playing football provocatively and their ball hit the ground far too close to us UN soldiers for comfort. In fact, the ball nearly hit me in the face, so I grabbed it and threw it away. Then I told the nearest Belgian, who had kicked the ball at me, that this mischief must end and that he and his comrades should go and play elsewhere. I could see he did not like that.

I was now looking at the same man with whom I had had the football dispute and he was our prisoner, having made a vain attempt to force his way through our roadblock, dressed in civilian clothes. He was taken aback when he saw me but I didn't move a muscle and just stared at him and his partner. During the interrogation we were exposed to a fresh attack from the Fouga. We lay down with the prisoners on the kitchen floor, putting our heads under a rickety wooden table – some shelter. The two prisoners, Michel Paucheum and Pierre van der Weger, were held in custody in a hastily prepared guardroom, and were well treated.

On 15 September, when the fighting was at its worst, a telegram arrived from Elisabethville for Cmdt Quinlan, asking 'Have you deserted your men?' What a dreadfully insolent question. As a result, Cmdt Quinlan flew into a rage and I can well understand why. We all felt sorry for him because he had certainly done whatever he could for the good of his troops in this crucial situation. Quinlan was a good officer with a strong fighting spirit. In spite of their aggressive spirit and indisputable courage, however, the Irish, Swedish and Indian reinforcements had not been able to get over the Lufira bridge where the Fouga jet had, of course, played an important role.

Saturday, 16 September brought a gleam of hope when, suddenly, a white UN helicopter came clattering along and began circling overhead. It seemed to be preparing for a landing but it was obvious that we were not the only ones who had noticed this fact. Then the shooting began, the likes of which I had never heard before. The air literally glowed with bullets as the enemy firing continued incessantly. Nevertheless, the helicopter continued its descent to a dangerously low altitude before landing – seemingly in slow motion – close to one of the buildings in our camp. It was a brilliant achievement. Then I got a happy surprise because both men in the helicopter were Scandinavians: the pilot, Bjarne Hovden, was Norwegian, and his mechanic, Erik Thors, was a Swede. We had scarcely had time to shake hands, however, before the gendarmes launched a strong attack in a fit of rage for not having been able to prevent the helicopter from landing. Whistling bullets made cracks in the wall just above our heads. The helicopter crew had arrived just in time because a couple of minutes later we heard the detested whining of the Fouga on its way to shoot down the helicopter, but too late.

That same afternoon, negotiations started with the Katangese in order to halt the bloodshed by calling a local ceasefire in Jadotville. As if by a miracle none of our men had been killed, although four or five were wounded. But the enemy had sustained heavy casualties. According to some calculations, their losses were from 150 to 250 gendarmes, including thirty whites. European civilians, who had stayed in Jadotville during the fighting, could draw their own conclusions from the fact that so many whites had bitten the dust. They saw that thirty coffins had been used during the funeral ceremony for those who died, and knew that only whites were buried in coffins there.

We were negotiating terms favourable to us, including the formation of an armed force of Irish and Katangese troops to patrol the area and, thus, restore and maintain law and order. We also sought an end to all hostilities. We demanded that the electricity and water be turned back on, that provisions, including beer, be delivered, and that curious onlookers be kept away. Everybody was happy the fighting was over. I could see that some gendarmes' eyes were bloodshot, while others were tearful. They had probably had a tougher time than we ever suspected. Later on, I heard from a civilian who had somehow got hold of 'inside information' that 1,000 gendarmes, or maybe more, had refused to attack us and deserted. I was not surprised because it must have been sheer hell each time they approached our positions. Many of the Irish soldiers were young, but they could certainly shoot.

We were sitting on the first floor of a building, drawing up the terms of an armistice, when we heard footsteps coming up the stairs. Who could it be? The door opened and in came a black gendarmerie captain. I stood up and asked him what he wanted. A minister had arrived, he said, and was waiting for us on the road outside. It came as quite a surprising blow to learn that the minister was none other than Katanga's Minister of the Interior, Godefroid Munongo. He was a brother of the chief of the Bayeke tribe and known locally as 'the only black man with a European brain'. In addition, Munongo was a grandson of Chief Msiri, the highly dangerous and powerful head of the Bayeke people. In the nineteenth century, Chief Msiri had fought against the Arab slave-traders. In 1891, he was murdered by a Belgian army officer.

Suddenly I had a brilliant idea. 'Ask him to come up here for a glass of cold beer,' I suggested. 'A minister shouldn't have to stand down there sweating in the heat.' The black man left. Comdt Quinlan looked at me puzzled and asked, 'What…?' I explained swiftly what I had in mind. I thought that we might capture Munongo by sticking a pistol in his back the moment he entered the room. With him as our hostage, we could get safe passage back to Elisabethville. Then we heard footsteps again. Could it be the minister? But no, it was the same captain telling us that Munongo didn't want to come up. He was too shrewd. The captain repeated that the minister wanted us to come and have a talk with him.

There wasn't much else we could do in that situation, so we went down. The minister was waiting for us, surrounded by a swarm of gendarmes. He behaved correctly and started by complimenting us for our brave resistance during the battle. I still remember what he said: we would be held in captivity in a place called Hôtel de l'Europe. He concluded his speech with these words: 'Je tiens à vous sauvegarder la vie' (I intend to safeguard your lives).

It was silly but the Katangese had already been allowed to enter our territory and, given the circumstances, a fight was not to be recommended. We had to swallow the bitter pill, no doubt, to avoid a massacre. Documents were drawn up and signed by Munongo and Cmdt Quinlan. Together with Munongo, I had already inspected the place where it was planned to hold us captive. The minister claimed there was running water, but I checked it anyhow by turning on a tap. He was right. The water shot out, splashing Munongo's elegant light grey suit. He accepted my apology, saying, 'ça ne fait rien' (it doesn't matter). In spite of the seriousness of the situation, I found it hard to keep myself from smiling.

We tried as long as possible to put off the time when we would have to leave the camp and really surrender to the enemy. We didn't move into the building intended for us until the afternoon of 18 September. During the trying period of captivity in Jadotville and Kolwezi, there were many nasty and dangerous incidents. Our first guards were black paratroopers who were well disciplined and we got on well together. Unfortunately, they were succeeded by gendarmes, and the difference between them and the 'paras' was striking. The gendarmes were morbidly suspicious of us, which caused problems. In addition, they were slacker and not as well behaved as their predecessors. Some of them may have had too much Simba beer – a 5 per cent brew. For instance, one of them was playing with his rifle as if it was a banjo or guitar, when suddenly it went off. He had obviously touched the trigger and the safety catch was not on. You should have seen his face – it really took him by surprise. When he thought about what had happened, he made a silly laugh. Luckily, the bullet hit the wall and nobody was hurt.

Now and then, gaudily dressed women came into the gendarmes' quarters. My friend, Lieutenant Joe Leech, counted our guards and estimated they numbered 90. Sometimes we had visitors: representatives from Union Minière and the Red Cross. But when a priest was sent to afford us consolation, it was a little disquieting.

One day, we even had a visit from President Tshombe himself. He didn't look too well and it didn't make him any happier having to wait quite a while for Cmdt Quinlan to appear. It wasn't easy for me to play the interpreter's role, trying to convince the president of Katanga that Comdt Quinlan would arrive at any moment. When he finally appeared, my position between the two men was not exactly enviable because they didn't like each other. There were journalists, photographers, some Irish officers and me in the middle trying to pour oil on troubled waters, so to speak. We were Tshombe's prisoners and anything could have happened.

Then a special commission came to inspect our conditions as prisoners of war. I knew some of the visitors very well. One of them was the commanding officer of the Swedish army's 10th Battalion, Colonel Anders Kjellgren, who was my brother-in-law. Another was Lieutenant Stig von Bayer, a Swahili interpreter and a good friend of mine from the 12th Swedish Battalion. The three Scandinavian captives – the helicopter crew and I – each got a bottle of Scotch, which embellished the day. We were all glad to see the visitors. Colonel Makito was also present and I heard later that he had been planning to kill all of us during the peace negotiations. I don't know if that story was true or not because one heard so many rumours.

After the commission had left, Colonel Kjellgren's loaded army pistol was found. I still don't know whether he had forgotten it – not very likely – or had left it for me, just in case. Anyhow, it would have been dangerous if the gendarmes had found it, with who knows what consequences for us. So, the pistol was immediately hidden.

One Sunday, the townspeople came up to our place of detention with their families to look at us, as if we were monkeys in a zoo. Meanwhile, the gendarmes were digging rectangular holes in the ground outside, which looked like graves. In addition, there were night-time raids on our quarters by armed gendarmes wearing helmets. They were nervous, afraid and utterly dangerous. As the interpreter, I was summoned at once to settle the situation as mildly and reassuringly as possible.

On one occasion, they suspected us of having a secret radio transmitter and confiscated our clotheslines on the roof of the hotel. I heard one Katangese soldier say, 'You never know with these white people, do you?' That was what they thought of us.

One day the deputy mayor, an African, offered to take me for a trip in his car. He said he would like to show me around the neighbourhood but first he had to do something, which wouldn't take long. Okay, I thought. The weather was fine and any change from the dullness of imprisonment was welcome. A few minutes later he drove into the gendarmerie camp, stopped the car and left me in it. So there I was, sitting alone, wearing my blue UN cap, which, for the gendarmes, was like a red rag to a bull. Then some people saw me, one of their enemies, in their own camp, alone. They pointed at the car and got very upset, shouting in loud voices, seized with frenzy at the sight of a UN man. They gathered round the car and some of them were bandaged, having been wounded in the recent battle against us. I locked the car doors as imperceptibly as possible, while trying to keep a stiff upper lip and appear unaffected. When they tried to open the doors, you can imagine how I felt. But where was the deputy mayor? At last, he was approaching – forcing out of the way those who tried to prevent him from reaching the car. He managed to get close and when I unlocked the door next to him he got in. 'Lock the door quickly,' I said, adding, 'Put the car in gear and full speed ahead.'. By then, some of the gendarmes had left the crowd in a hurry, probably to get their rifles. If they had been able to pull me out of the car, I am quite sure my fate would have been sealed. In any case, our excursion was off – too close to a lynching, my own.

On 13 October, we were moved by bus from Jadotville to Kolwezi. On the way there the slopes were rather steep and at least one slope proved to be too much for the old buses, so we were ordered to get out and push them uphill. With a united effort it worked. After that there was a break and some of the gendarmes climbed up on top of the buses with their carbines to keep an eye on us and prevent any attempt to escape. Some other gendarmes left the road and walked down to a little river, which was bordered by trees. Howling and guffawing, they started shooting at the poor, scared little monkeys up in the trees. The monkeys fell dead to the ground, looking like tiny babies. Then the gendarmes proudly returned to the buses with their quarry. The monkeys would certainly constitute a good meal later, but I will never forget the sight of these 'brave' armed hunters… so-called human beings.

We continued on our way towards Kolwezi but the drivers must have made a mistake because when the buses stopped we were not in Kolwezi at all but near a gendarmes' training camp. The camp was near a lake or watercourse and there were a lot of rather young people, men and women, there. I got off the first bus and the Irish troops also left the buses. The Africans were very curious and looked quite happy. A young woman looked at me and then bent down to pluck a handful of grass and threw it over her shoulder. I saw a paratrooper in his red beret and asked him what the woman's gesture meant. 'It means that somebody will die,' he answered. And when I asked him why they all looked so happy he said they hadn't seen meat for three weeks.

Then the black district-commissary came running and was very upset. We had to go to Kolwezi at once, he said. We really didn't mind. I shouted to my Irish friends to get into the buses again and told the driver of our vehicle to try to find the right way to Kolwezi this time. Back on the dusty roads, you couldn't see your hand in front of your face. Clouds of dust whirled in the air and came through the broken windows, so the front of my uniform turned quite red. I covered my face with my shirt in order to prevent my lungs filling with dust and getting an attack of coughing.

Our eventual arrival at Kolwezi, in darkness, was one of the most frightful and ill-boding events during my time in Katanga. Nervous, hateful and fanatical 'darkies' in steel helmets – and careless with their cocked rifles – were our new prison warders. They seemed brutal and even more suspicious than the gendarmes in Jadotville had been. Given their behaviour, they must have heard terrible stories about our 'sweep' there. When the first bus stopped, the door was jerked open and one of the Africans climbed aboard uttering some guttural words and thrusting his rifle into the chest of my friend Bill Donnelly. As if that wasn't bad enough, a black captain, wearing spectacles, pushed the Katangese soldier aside, wrenched his rifle from him and howled something like 'This is not the way to treat these types.' While roaring this, he hit Bill Donnelly a furious blow in the stomach as hard as he could. Bill just sank down on his seat, unconscious, I think.

All this happened very fast and as soon as I could, I got out of the bus. Immediately, I was shadowed by a guard pointing his rifle at my back. The Irish soldiers were then told to take off their caps and shoes, and a meticulous check of our luggage began. A black lieutenant started to hit two Irish corporals hard in the face. They were standing to attention side by side, not saying a word. I was sure they would remember him and pay him back at the first opportunity. When I warned a black officer that it certainly would lead to serious trouble if this brutal treatment didn't stop at once, he just screamed at me: 'Lieutenant, je ne l'accepte pas' (Lieutenant, I don't accept it).

This maltreatment and the rough treatment some of us were exposed to was due to the fact that some cartridges had been found in a part of our luggage which the owners hadn't seen during the three weeks they had been prisoners. They themselves had no idea that there was any ammunition in the luggage. Besides, what could you do with a few cartridges when you didn't have a single firearm and the place was swarming with guards? Oh yes, there was a weapon – a little, home-made catapult which one of the younger Irish soldiers had amused himself with. I was later informed that it was dangerous. O, sancta simplicitas!

After a lively debate with the Katangese, we agreed on how to proceed: an Irish officer would check that there were no weapons or ammunition, but if there was anything like that, it was to be handed over to the blacks. The same evening, I reported the maltreatment that had occurred to the local district commissary, who proved to be an honest and judicious man. He made it very clear to the blacks that no cruel or humiliating treatment would be accepted. He told us that we would get all we needed, and he kept his word. The following day we got a nice man, a black captain, as our new chief prison warder. He took care of all that concerned us. He was a real gentleman and we all liked him.

During our stay in Kolwezi, the same African lieutenant distinguished himself by his sadism. Earlier in Jadotville, some Italian orderlies had increased the number of prisoners of war. Now a couple of Irish prisoners – but not from our company – had come to Kolwezi. They were not allowed to stay with us, however, because they had been captured near Camp Massart in Elisabethville in civilian clothes and were, therefore, regarded more as spies than POWs. Somebody told me that they had been taken by the sadistic black lieutenant into a nearby building and ordered to clean a toilet. They were locked in the toilet but were not provided with any cleaning materials. It was a mean and insulting attempt to make them protest, which would give the lieutenant an excuse to punish them one way or the other – e.g. by whipping or other torture – or perhaps shoot them. I ran into the building and found a gendarmerie major. He was Italian, although he said that he was, and felt like, a Katangese. I was very upset and demanded that this disgusting treatment had to be stopped immediately, which it then was.

During our period of imprisonment in Jadotville and Kolwezi, we did what we could to cheer up the men and, to this end, different activities took place. I myself took the opportunity to teach some French and I also gave talks about Sweden and Swedish girls. The latter subject, I know, inspired my young audience with enthusiasm!

One day, the Irish Army chaplain, Fr Fagan, came up to me wondering if I could help him with a double wedding ceremony in the prison camp. Two black corporals – they were signallers – had asked the priest if he would help them to get married according to the rites of the Roman Catholic Church. But there was a snag: they had already been married by an American Methodist pastor. Their reason for wishing to be married once again, this time in the Catholic tradition, was that they wanted to be 'properly' married. Fr Fagan said he would be glad to help them if the Methodist pastor had no objection, and if they received permission from their army superiors. So, at 10 o'clock one Sunday morning, I performed as a kind of assistant priest at this double wedding. As the Africans didn't understand English, French had to be used and that's where I could be of some help. I stood before the two brides and two grooms asking the right questions in French. Behind me stood Fr Fagan in full canonicals, while the men of A-Company kneeled before us. It was all very solemn. After the affirmative answers of the participants, Fr Fagan sprinkled them all with holy water and even christened the infant of one of the married couples. They were each presented with a set of rosary beads, after which a wedding breakfast was arranged.

On 16 October, we were brought to Elisabethville for an exchange of prisoners that proved to be abortive. On our way there we were delayed and had to spend the evening and part of the night in Jadotville. At 1 o'clock in the morning we were woken up and ordered to prepare to go to Elisabethville. At about 9 o'clock in the morning we arrived at Camp Massart. Our stay there was a most unpleasant experience; sitting in the buses, while the hot air nearly cooked us in the merciless sun. We spent all that day in the buses with no food or drink. Finally, without any explanation, we were transported on the long journey back to Kolwezi, with a short break en route at Jadotville, arriving at about 4 o'clock in the morning. Not until then did we get something to eat.

The next time we were brought to Elisabethville, it was a very nasty and dramatic experience. On the afternoon of 25 October, we were back in the capital of Katanga where the planned exchange of prisoners finally took place. We were free again – what a wonderful feeling! At the same time, the UN released forty-five Katangese prisoners. As I was celebrating our release, something unexpected happened. No sooner was I back in Swedish Army service than an Irish soldier hurried up to me and asked if I could come over to talk to Commandant Quinlan. 'Of course', I answered.

I found Quinlan together with President Tshombe, and their attendants, both black and white, and, of course, members of the press. I first turned to Quinlan and asked, 'What do you want me to say to the President?' 'Whatever you like, Larry,' Quinlan answered. So I turned to the President and asked him in French: 'Monsieur le Président, est-ce que vous avez peut-être des questions?' (Do you have any questions, Mr. President?). Tshombe grinned derisively and asked why there were so many UN prisoners and so few Katangese prisoners. Then I asked him: 'Do you really want to know?' 'Yes', he said, still grinning. 'Well,' I said, 'Go to Jadotville and dig in the ground. There you will certainly find your brave soldiers.' His grin disappeared. Then I asked him, 'Encore des questions, Monsieur le Président? Sinon, au revoir, Monsieur le Président' (Any more questions, Mr President? If not, farewell Mr President).

Commandant Quinlan looked very pleased. I hadn't been very polite while talking to Tshombe, I'm afraid, but I didn't like his way of trying to make fun of Quinlan, who didn't speak French. It wasn't gentlemanly. The men of A-Company and myself had been held hostage from 18 September to 25 October. We were Tshombe's trump card in his truce negotiations with the United Nations.

17

A Retired Chief of Staff Reflects
on the Congo

Company commander Louis Hogan served with the 33rd Battalion in the Congo. In later years, he rose to the top, serving as Chief of Staff of the Defence Forces from 1981 to 1984, with the rank of Lieutenant-General.

The year 1960 was a flattish year – it started off that way – and then suddenly the Congo burst open. It was known that we were going to be asked for troops, and we were. We were asked for a lot of troops, and in fact we had two battalions in the Congo at any one time. The strength of a battalion varied from 700–800 men, which was a big contingent for a small country. The first two battalions were made up of four companies each.

Col Bunworth was my commander, and I was the company commander under him – in charge of A Company. As a Protestant in the Irish Army, Bunworth was quite a rarity but his brother was also in the Army. His family had been very active on the republican side in the War of Independence so the Government did not forget them when it came to selecting men for the Congo. I think Bunworth was a Cork man.

The situation in the Congo in July 1960, just after independence, was very unstable and quite dangerous. The Congolese had sent a team to a big conference in Belgium [in January 1960]. When the Congolese got there they found that they were way behind the other [African] nations which had outpaced them. So, when they went back they started to agitate and to start trouble. That is what started the situation in the Congo. The powers at the time, like the Americans, Belgians and English, were all interested in the resources of the Congo. They did not want to get out but the Congolese wanted to get them out. The trouble started then. It is alleged that at one stage there were something like 22,000 UN troops in the Congo, which incidentally were commanded by an Irish officer, Lt Gen Seán McKeown from County Louth.

I did not get the impression that the Congolese were capable of running their own country. For instance, in the town where I was based, Albertville, which was on the shores of Lake Tanganyika, the Congolese took over quite a big store. The locals, who had been there for years, came in to buy stuff. They bought and paid for it until the day came when the store ran out of stuff. The

18. Former Irish Army Chief of Staff Lt-
General Louis Hogan

Congolese believed then that they had been ripped off – that the guys inside had put the money in their pockets, whereas they hadn't.

One of the dangers we faced was that our fellows would get too friendly with the Congolese. We had to be neutral between the Congolese and the Belgians who were withdrawing. Of course, when you are withdrawing you are vulnerable so we had to be careful at that stage. I take the point that having been engaged in an ex-colonial war ourselves we would be sympathetic [to the Congolese] and that was one of the things we had to watch. But then the Congolese themselves did a lot of damage to their own cause insofar as they fired on UN troops in many places and killed them. One can see that in the history of Niemba where nine men were killed. That was my company. I knew all the men who died. The trouble with Niemba, and with many things, is that one of the paragraphs of the agreement between the Congolese and the United Nations specified that the UN would have freedom of movement. In other words, they could move where they liked, when they liked without any hindrance from the Congolese. A patrol went out one day, south of Niemba on a circular route, and they came across this [damaged] bridge which effectively blocked the road, which was against the rules. That is how the situation arose.

We did come across atrocities but I would prefer not talk about them. When I first heard about Niemba, I was in Albertville. The job my company had was guarding the airport there. It was quite a big airport, maybe thirty miles from Niemba. In the follow up operation after the Niemba ambush troops were sent from Albertville to try to find the survivors. After the fighting at Niemba the garrison there was withdrawn to Albertville. There were one or two missing and a party was sent out to search for them. I was involved in it insofar as my troops were sent out but I did not go myself. Colonel P. D. Hogan went. If I had got the opportunity or offer of going I would have gone, but Col Hogan was Col Bunworth's No. 2 and he had the information [about the Niemba incident] before I had. He arranged to go himself with his party.

It was one of the worst things that ever happened in the Irish Army, although many others were killed also in Elisabethville in the south when they were trying to get rid of Tshombe. Eventually a fight started there. While as a policy I would say 'no', it could happen that the Balubas could mistake our troops for Belgians. Troops go out and they [the locals] see white men with weapons, so it could happen. One of the interesting things about it was that the leader of the Balubas at the time was an ex-Sergeant Major in the Belgian Army. It is very hard to train soldiers to soldier properly – they do not concentrate, they do not listen, and they do not apply it – but there are a couple of things that you must do in soldiering. You must spread out. You should not be in a bunch because if you are in a bunch a machine gun can get a lot. If you spread out, however, they cannot touch you. This ex-Sergeant Major, who was head of the Balubas, got the Baluba army and trained them quite simply – brilliantly actually. Firstly, he said: 'Now, the white man's – as they referred to us – guns will not touch you if you do what I tell you.' They nodded agreement and said 'OK'. Then he made up a potion of all sorts of old leaves, drugs and everything, and gave a drop of this to every fellow to rub on his chest. Then he said: 'If you touch the man next to you the magic will disappear from that potion,' so straight away he had his dispersal because his troops would not stay close together. Secondly, he said: 'If you turn your backs to the white man and run away the potion will not work either because the magic disappears from it.' We cooperated with that to the extent that our instructions at the time were that if we had to fire at the Balubas or the Congolese, we would fire at their feet or legs. That then showed the Balubas that our friend [the Baluba leader] was right because shooting a man's legs is no game. It was interesting. I learned about this Baluba leader from the Belgians who had a lot of troops there. We associated with the Belgians but not in a friendly way. The instructions I gave to my company were to behave towards the Belgians in a reserved and friendly fashion – I mean, you met them and would speak to them. A lot of people do not agree with me, but I must say that the Belgians treated us very well. I think they did it much better than we would do it here if the positions were reversed. They would give you tips like that, you see. Although I was not told personally, it was probably Belgian army officers who told the story about the ex-Sergeant Major who trained the Baluba army. I never met the Baluba chief. He could well have been involved in Niemba but if he was he escaped because eight or nine fellows of the [Baluba] party at Niemba were arrested. We flew down one night from Albertville to Manono where they were [in hospital]. I was there and the next morning we arrested them before daylight. We whisked them into the plane and we were gone. That was Operation Shamrock. It was not done in contravention of UN regulations but a lot of people had reservations about it – going into a hospital, pulling out wounded fellows, putting them on a plane and taking them away. There were reservations about that, yes. Very, very few people knew about it. I was at the conference at which it was planned. It was purely an [Irish] army planning conference. In the end the Balubas concerned received very light prison terms, if at all, because they had to be handed over to the local authorities. Of course, the local authorities were not going to give heavy sentences to their own people who were fighting the white man. They, allegedly, got prison sentences of two and three years, if at all. They could have got a heavy sentence but did they serve them? We had no way of knowing. It would have been a cause of disappointment amongst fellows in the Army. Another battalion went out immediately after us and they had to fight. Soldiers get very logical and down to earth about these things. Okay, you know how to look after yourself, you see — the other thing is past and gone. But when the word came back that night, the troops were very angry. The troops themselves wanted to go out and arrest these [Baluba] fellows. Our troops did not have to be restrained from doing so, however.

Looking back at it, the UN presence in the Congo served a purpose because it saved the authorities, let's say in Belgium or New York, from doing anything. You are there and you are not doing anything very much. Yes, it did serve a purpose. As a matter of fact the UN troops are so popular now, and so much required, that they are scattered nearly all over the globe.

18

The Army Doctor's Story

*Dr Joseph Laffan served as a colonel with the Army Medical Corps in
the Congo. Here he recounts his experiences during those early days
of UN peacekeeping work.*

In 1959 I was supposed to go to the United States for a conference, as a result of
which I had to get a smallpox vaccination which was crucial in a way. In June
1960 I was on a few days leave down the country. I got a phone call one morning from army headquarters to say: 'Will you go to the Congo? We want somebody to go there.' One of the reasons I was picked was that while I was the right
age and readily available, I had had my smallpox vaccination and that meant I
could go quickly. I agreed once my wife consented – that was on a Wednesday
morning. I went back to Dublin and the next couple of days were very busy getting everything ready. On the Saturday morning myself and another officer,
Comdt Joe Adams, left by commercial airlines for the Congo; first to Paris, then
to Marseilles and from there to Brazzaville where we disembarked and crossed
the river by ferry to the Congo. We reported to the chief of staff of the UN mission in the Congo, General von Horn. Joe Adams and myself were the advance
party to make things ready for the battalion that was to be sent out as soon as it
was assembled. Everything was done in a great hurry with ad hoc arrangements.
For several days we worked in Léopoldville on General von Horn's staff. Other
officers and NCOs came out later, so there were about seven there. Some five
days later we were sent up country. During that period we were all doing various jobs. Joe Adams was doing operations and, although I was a doctor, I was
doing intelligence and quarter-mastering. We had to get things done; trying to
arrange supplies and getting such intelligence as I could from Belgians and other
sources. Fortunately, I had done a command staff course some years earlier so
that a lot of these things were not altogether strange to me as a medical officer.

I received instructions from General von Horn that I was to go to Kasai
province, but he changed the instructions overnight, so the next day I was sent
to Kivu. We left from Léopoldville at dawn and flew across country in a DC-4
aircraft with a load of stuff, including cans of petrol. We had no medical supplies, we were poorly equipped in that way. The night before, I had gone to the
kitchen of the hotel where we were to gets packages of sandwiches and bottles
of Coke. They were the only provisions we had when we were leaving

Léopoldville. We flew to Kindu on the river, it was the hottest place I've ever been in my life. We landed there and met some Swedish UN troops who were already in position. One of our officers, Comdt Pat Liddy, who happened to be the battalion's legal officer, and a sergeant, were left there to prepare for the arrival of A-Company. The rest of us flew on to Goma in Kivu province and landed at an airfield beside a big lake. The Congolese did not know who we were and they were fiercely suspicious. We were arrested and taken off in trucks to be interrogated. All our stuff was confiscated. Fortunately, we had an interpreter with us – a Swedish officer who had lived there and knew the local language. We were interrogated by the [Congolese] army first, which was rather interesting. We were in a small room with a couple of windows through which we could see women and children looking at these strange white men. We told them of our circumstances and our mission, but it was not easy to convince them. We got a fright during the interrogation when a message came to them that more troops were landing at the airfield. We did not know who in the name of God these people were because none were scheduled to arrive. Adams asked me to go down to the airfield with the Congolese to see what was happening. A Congolese soldier drove me to the airfield and I found that a small private aircraft, flown by a Belgian, had come in from Rwanda. That incident was settled quickly and it turned out to be a fortunate one because on the way back I passed two large buildings. I asked the driver to stop and discovered they were two schools that had been evacuated. They looked quite suitable for our headquarters. I went back to where the others were detained and told them what I had found. Then the Congolese softened a bit and we were put back in the lorries and taken to a civil administrator named Pierre Rwangi, the Congolese district commissioner who had only been in office a few days. We got on well with him so a deal was stuck whereby the Congolese would co-operate with the battalion that was arriving the next morning, we could have the use of the two buildings I had found, and they would return all our material, including Congolese money in one of the boxes. As a sweetener, I offered to look after their illnesses when we got set up.

There was a Belgian resort hotel a short distance away where we got rooms for the night. We arranged to have a dinner there with the Congolese military and the civil administrator to seal the pact. A while later the Congolese trooped in, but the Belgian residents in the hotel were horrified to see these blacks coming in. We were all in uniform and we could feel them glowering at us. The conversation was exceedingly difficult during the meal because some of them knew a bit of French but none knew any English. We were struggling with the little French we had. I was sitting opposite the military commander who had been an NCO a week or two earlier. He was sitting there with a grenade hanging by a safety pin from the lapel of his jacket. There was a bodyguard behind him. After the meal we went to bed jaded because we had been up since four o'clock that morning. We were awoken at four o'clock again and brought to the airfield. We were expecting Americans to arrive with elements of the 32nd Battalion. We had no signalling arrangements of any sort to give a message to the incoming flights. We arranged that those of us with blue berets would stand along one side of the runway. We hoped we would be spotted by the incoming aircraft. As dawn broke I was surprised to see two big volcanoes nearby, one of which was active. Sometime after dawn an aircraft appeared flying low over the runway and we must have been spotted as friendly troops because it landed soon after. When the troops disembarked I met the commanding officer, Col Mort Buckley. We had arranged for the Congolese to bring along some trucks and drivers to transport our troops. Soon after another plane arrived. We then ferried the troops to the empty school buildings I had discovered the day before. We set up our local headquarters there and had a meal of bully beef, tea and hard biscuits like pieces of wood, but we were glad of that meal.

The unfortunate private soldiers were all volunteers and none of them had any previous overseas experience. Many of them did not even know where the Congo was. They had all been vaccinated against smallpox which was a require-ment, as a result of which many of them had high temperatures and were sick. There had been no time to get tropical uniforms for them. I had scrounged some tropical uniforms in Léopoldville from Belgian sources, but not enough. They were Belgian military tropical uniforms which a few of our men wore. I signed for things there, I don't know if anybody ever paid for them but I got them.

I could speak a little bit of French so I went off in a truck to the border with Rwanda. I was able to tell the driver where to stop as we set up the outposts along the border. Over the next few days we were able to hire transport of our own, including a number of Volkswagens as staff cars. The troops moved in and made themselves fairly comfortable. Our mission was a political one based near the centre of Goma which was five or six miles from the Rwandan border. Patrols were sent out to various places to establish the position, show ourselves and show that we were a force. The headquarters company and A-Company were in Goma. One company dropped off at Kindu, while another was estab-lished a little bit later about 100 miles south of us at a place called Bukavu, at the southern end of the lake. We were scattered over three locations in the province, but we were spread very thinly.

Routine patrols went here and there, wherever they were required and I went on a number of them. We went quite a distance north on occasion and twice I went with them into Uganda. We established fairly good relations with the administration in Uganda which was about 100 miles away. We had no problem in crossing the border, it was done by arrangement and it was not heavily defended. You would scarcely notice that you were crossing it. We were not very far from the Equator so a number of our patrols going north would cross it and that was of great interest and a fascination for the troops. They would hop off the vehicles and walk across the Equator and photograph themselves there. Our patrols went up as far as Beni where we established contact with some Irish Red Cross workers who had a station there looking after lepers. The town of Beni was in the foothills of Mount Ruwenzori. It was a large mountain just

19. Dr Joe Laffan of the Army Medical Corps in
Goma, Congo, August 1960

above the Equator, which was permanently snow capped. There are a lot of volcanoes around there too, in the Rift Valley area. Some of them were active and
you could see recent lava flows. The most recent ones were black, while others
were dark gray or light gray with little bits of vegetation growing on them. It
was fascinating to see the lava flows.

Within a few days I set up a dispensary and each morning I saw Congolese
patients and tried to do what I could for them. That was a good public relations
effort because it helped to have us well accepted. They had had a medical service before that but the Belgians had gone. Some West Germans took over the
running of an abandoned Belgian hospital and we had good liaison with them.
We used to protect them with armed guards when necessary. There was no
fighting in the area, although we had to treat accidental gunshot wounds suffered by Irish troops.

There were very few cases of syphilis. I had plenty of experience of that during the war. There was some gonorrhea but not much. There were not many
opportunities for the men. Some of them consorted with the black women who
were eager enough. It is unlikely that they were prostitutes.

The 33rd Battalion came out later than us and set up in Albertville. Their
men were involved in the Niemba incident. I was there the day it happened. Col
Buckley wanted to have a liaison with Col Bunworth the commander of the
33rd Battalion. We had our own aircraft at that stage, a small Otter. He decided
to fly down to Albertville and asked me to accompany him. It was a rather hairy

flight along the edge of Lake Tanganyika. There were some storms and I was a bit anxious about it because there were cliffs on one side of the lake so there was no place to land. It was rather awkward. We met the headquarters staff in Albertville and had lunch with them. I met my opposite number, the medical officer, who was Comdt James Burke, generally known as 'the Badger'. That was the day [8 November 1960] the ambush happened at Niemba. Later in the day we flew back to Goma, unaware that during those hours the men had been killed at Niemba. The news hadn't got back. The next morning we heard via our radio communications that these men had been killed. Col Buckley was sorry for his opposite number, Col Bunworth, so he gave him his aircraft which was very handy for getting around to the more distant places. There was a feeling of horror and dismay when the news came in about Niemba. Up to that time, as a non-combatant, my only weapon was a 9mm pistol which I carried in my pocket. I did not bother with a holster. After Niemba, I was told, 'You'll get a sub-machine gun.' I did a short training course with the Gustav which was my personal weapon until I landed back in Dublin at the end of the mission when I handed it over. Everywhere I went in the Congo I carried that personal weapon like the combat soldiers, but I never had to use it. We felt we couldn't trust the people that we thought we might have earlier. And, from the other point of view, if you negotiate anything it is better to do so from a position of strength rather than from apparent weakness. If you seem to be formidable and armed you can be treated with a little more respect, or a lot more respect. We made sure that happened in practice. I felt there was something contradictory in a doctor carrying a submachine gun but the alternative was worse. One of the men who was killed at Niemba was a medical orderly named Farrell, from the hospital where I worked, so that hit hard. There is a little monument to him outside St Bricin's Hospital. After that, all medics were armed.

There were no Balubas in Kivu where we were, it was a different tribe. They were very agreeable and we got on well with them and liked them. Many of the locals spoke Swahili which was a *lingua franca*. It is a mixture of Arab and African dialects. There were local languages also. Some of them could speak French, but very few of the Irish could speak it. One evening I met a young chap who proudly showed me a book. He was studying Latin through Swahili, neither was his own language. He may have been an altar boy or something like that. We had our own chaplain, Fr Crean, who became parish priest of Donnybrook after returning from Africa. He never carried a weapon. He would say Mass every morning soon after dawn, and the Rosary in the evening. The soldiers were more devout then than they would be now.

Not long after that a decision was made to move the battalion down to the Kamina base in Katanga. Our place was taken over by a Nigerian battalion. We flew down to the huge airfield at Kamina and had a different regime there. It was like a Belgian town and we had a fine officers' mess and a swimming pool. The Belgians looked after themselves very well. We stayed there until January 1961. Many of the Belgians were in fear of their lives and a lot of them had fled.

I remember meeting one poor devil who had spent about twenty-eight years there. He fled with only a couple of francs in his pocket. It was hard for people like that who left everything behind, including a farm and villa. There was a fully established Belgian colony in Rwanda with all sorts of people there. A small number of Belgians stayed behind for a while but they gradually went away. There was a fine Belgian doctor called Coutelier who stayed on for some weeks. I used to meet him fairly frequently as he did the surgery in the local hospital. Somehow or other, however, he and some others were either told to go or felt they should go.

For a while there was nobody in the local hospital but one day after the Belgians had gone I received an urgent call to come there. An unfortunate African had been attacked by somebody with a spear, probably in a domestic row. He had a big abdominal wound and about half a yard of intestine hanging out. There were no medical staff apart from myself and while I had never worked as a surgeon I had done a lot of assistant surgery, handling anaesthetics. My main field was pathology but I often used to assist during operations in army hospitals. Because there was nobody else to do it, I took it upon myself to deal with the African's injuries and he recovered. I was lucky. I gave him an intravenous anaesthetic, pentathal, with little increments after that to keep him under. I cleaned the bowel, reinserted it, stitched up the peritoneum and the muscle as you would for an appendicectomy. I gave him such antibiotics as I had available and he survived. I was rather pleased with that, being an amateur surgeon.

I used to go up to Beni to visit the two Irish doctors at the Red Cross unit there. They were looking after a leper colony and we visited them twice. One of the doctors later became a skin specialist in the Mater Hospital in Dublin, and the other was called Moloney. The first time I went there it was as part of a four-day patrol. We visited various places along the way, including Butembo where bubonic plague was still endemic. On that occasion I just missed an earthquake by twenty-four hours, which was rather lucky.

There was no fighting in the area I was in. There was only a small interregnum between the departure of the Belgians and the arrival of the Irish. The Congolese soldiers stayed on in the barracks and other posts they had. Some of them promoted themselves to officers and commanders, which was fair enough. We tried to keep good relations with them. On two occasions we visited their headquarters at a place called Rumingabo. I felt a bit unhappy about it, however, because we did not know exactly how they were going to react at any time. We were on a political mission to keep things quiet, keep in contact with everybody, try to find out what was going on, and keep the peace. We had constantly to send reports to headquarters. We did that fairly well in the areas we were able to control.

I will show one way in which I noticed that the locals appreciated the difference between the Belgians and the Irish. One day, I was on a patrol and came to a village I had never seen before. Apparently the people there did not know about us. When they saw us coming along they all bowed down putting their heads on the ground as if they were terrified. That gave me an insight as to how they felt towards the Belgians. They thought we were Belgians so they bowed down with their foreheads on the ground. They did not know who we were or what we were, but they saw these armed white men coming along. The blue beret meant nothing to them. It was hard for the locals to distinguish between the Belgians and the UN troops. It took an effort to get them acquainted with us. That was our first visit to that big village. I got the impression that Belgian rule had been severe, especially the places where they weren't in overwhelming strength but would just come to occasionally. But I did not see any evidence that the Belgians had acted brutally.

A certain amount of commerce still went on by boat or long distance refrigerated lorries. Once we even got some rainbow trout which must have been frozen from a fish farm. The food was all right but not great. On another trip we called to a Cistercian monastery, north of Goma, in the middle of a tea plantation. We visited the Belgian monks who were very pleased to see us. They owned and worked the plantation which was a money-spinner for their charitable works. They must have had excellent relations with the people in the vicinity. There were cases of religious being slaughtered in the Congo, but not around there.

We were very busy during the first few weeks and then Sunday came and things were a little bit slack. I thought I would go for a walk to a small volcano which was nearby. I had my pistol in my pocket, hoping that I would not meet anything dangerous as I was by myself. When I came to the top of the volcano's crater I saw burnished, shiny basalt lava which had congealed after overflowing. There on top was an empty Sweet Afton packet – one of the others had been up there before me!

We had good relations with the local people in other ways too. There were camp followers, little boys, many of whom were pretty ragged and poor. They would hang around the camp and get friendly with the cooks. Often enough they got a hand out. Some kindly NCOs organised a subscription towards buying a little jacket, shirt and shoes to deck out some of the little lads in a nice clean outfit. They were so proud of it. The local tailors made these children's outfits. In fact, I got khaki shorts for myself made by a local tailor.

The UN presence gave the civil administration a chance to get organised, until they became too corrupt. Their corruption made things as bad as, or worse than, ever. When the UN decided to go in there had been some massacres and nuns had been raped. It looked as if the country might fall into chaos. There were very few educated Congolese. As far as I know there was only one Congolese university graduate at that stage, or so I was told. So, they had not much material with which to form the basis of a sound administration. I felt that was a reflection on Belgian rule. Given our history, the Irish in the Congo would have been very much pro-Congolese and anti-colonial. Many good officers served there, both with the battalions and at headquarters in Elisabethville and Léopoldville.

I did not meet any mercenaries there. They came a bit later during the Katangese secession. Then the trouble really began, but by that time I had returned home. I was glad to get home for family reasons. The post had been difficult and slow, which was a bad thing for the troops who were not accustomed to the idea of being away from their home country, like the British or French might be. When the mail arrived there would be rejoicing. When we got the sacks of post we would put it on a big billiard table and sort it out. There would be jubilation when the troops got their letters.

I made a trip to Léopoldville once and we were told that some post had arrived. It was locked up because it was a Sunday so we could not bring the bags back. It was days later before we got them. I had received word in a letter from my wife that my youngest son was sick with rheumatic fever. It had been going on for a long time and I was very worried about him. We had no direct communications but one evening I mentioned the matter to our radio operator who said he could do something for me. He called up a ham radio operator in Uganda and gave him a message. That operator called up a Fr. Stone, a Jesuit at Rathfarnham College who was also a ham radio operator. Fr. Stone picked up the phone and called my wife, saying: 'I've got an inquiry from your husband about your son. Can you tell me what the position is?' Fr. Stone then called Uganda again and I got the message within an hour from our radio operator. The news was fairly good and it had been ten days since I had heard anything, so I was relieved.

There were funny incidents, too. We had one cook who scalded himself with a teapot. He didn't want to admit that he went to the Congo and had that kind of accident, so he wrote home to say he had been injured in a grenade explosion. The word went out to army headquarters that somebody had been injured in a grenade explosion. We got a rocket back quickly from HQ seeking details of the incident. When I investigated the matter I found the cook had been exaggerating. Soldiers might do silly things like that – making things look a bit better when they wrote letters home. The grenade incident was not accepted officially but the story had circulated in Dublin and HQ were surprised because they had not been informed of it; not surprisingly, because it never happened.

There were banana plantations in the area. The women used to do a lot of the work cultivating the fields, while you would see the men hanging around smoking or gossiping. Sometimes you would see the women collecting big bundles of sticks which they carried on their backs for firewood. We used to buy lots of bananas from the locals.

I had a chameleon as a pet when I was there. He lived in a bush in a pot outside my medical hut. I used to call him Ferdie; he was named after a senior officer who shall remain nameless. I got the chameleon from one of the locals. I fed him insects which I would catch and hand to him. His tongue would shoot out and grab them.

There were two other army medical corps doctors there at the time. Edward Daly was in Bukavu which was 120 miles away, and Henry O'Shea was in

Kindu, about 250 miles away. All of us had been recruited at short notice for service in the Congo. We did not have sufficient medical supplies there. We didn't have much time to collect things and did not know what we needed. When I had been out there a short time I found a number of deficiencies so I bought some medical supplies locally and scrounged others. After a while the UN developed its own supply system but that was slow and difficult. Medical supplies had to be requisitioned through a chain of command and one would hope to get them. Later units were better equipped having been advised what to bring with them. For example, there were no anti-tetanus doses or things of that nature. The troops received anti-smallpox vaccinations before travelling to the Congo, which made them feel sick and miserable with a high temperature, sweating. I had that reaction myself when I got it the first time. It was brutal to take these poor devils off a plane and send them off for duty straight away. The officers had much the same experience because very few had received prior vaccinations, as I had.

In theory there was a class division between the officers and men, but in practice in the Congo it didn't work like that because we were all in the same boat. Rations were very much the same all round and everyone had the same needs; they led a rather isolated life. The days were busy and there were not many comforts. There was no bar even though we were able to get drink, but not a lot. I was keenly conscious of the question of morale and we tried to do what we could to provide morale boosters, including bingo games. Powdered milk and chocolate were delivered to the camp, but all the inscriptions were in Russian. I had studied Russian and was able to figure out that these supplies had been sent from the Soviet Union to their embassy in Albania. Somehow or other it all landed in our camp, so Russian chocolate intended for the Albanians ended up as bingo prizes!

We also organised table tennis tournaments in the evenings, using a covered billiard table, to try to boost morale. We had no real rest and recreation facilities. On one occasion I saw an officer who was under stress and I was able to get him a few days off; it was to keep people from breaking down. I felt there was a risk of breakdown, although it did not come to that because we all became aware of it and tried to be supportive. We tried to organise diversions but the scope wasn't great. The risk of breakdown came from boredom, separation from home and anxiety. A lot of these troops were youngsters who had never been outside Ireland before. At that stage continental holidays were very rare. I remember we had one fellow who thought he was going to Belgium because it was called the Belgian Congo. One fellow did break down on the plane with a later battalion. He walked to the Congo – walking up and down all the time aboard the big globemaster aircraft. The plane was filled with canvas bucket seats so it was referred to as travelling 'hard arse'. It was not very comfortable for a ten-hour flight.

19

An Army Chaplain Remembers

Fr Ronald Neville was a chaplain with the 37th and 39th Battalions in the Congo, from 1962 to 1963. Here he looks back at his time in Africa, forty years ago.

Early in 1962, I stood in for Fr Cyril Crean, who was head chaplain to the Defence Forces and based in Collins Barracks, Dublin. I was just three years a priest and only a few months a temporary army chaplain. The acting head chaplain, Fr Pat Duffy of McKee Barracks, asked me if I would go to the Congo with the 37th Battalion. I was delighted to oblige.

Everything was delightfully new to me. In those days there was no learning course for new chaplains. A commandant in Brigade HQ, next door to the chaplain's house in Collins Barracks, showed me how to give the salute in the march past of the battalions in McKee Barracks. Afterwards, I got a private reprimand from a staff officer at GHQ for not keeping my left arm stiff by my side while I saluted with my right hand as I marched past the reviewing stand.

The inoculations sickened me but I recovered while on a brief period of leave at home to say goodbye to my parents. It was not so long since the coffins of soldiers killed in the Congo had proceeded through the crowded streets of the capital. I was with all the officers of the battalions when President de Valera gave a reception for them at Áras an Uachtaráin.

Prior to our departure from Baldonnel, I visited the soldiers and gave them the opportunity of going to confession. Once you got the first few to come, nearly all the others were glad to follow.

My next recollection is that of sitting uncomfortably on a wooden seat that went along the side of the American army transport plane, a Globemaster. Luggage and equipment were stacked in front of us in the centre of the plane. Our return home and all subsequent overseas journeys were by civilian transport.

Previous Irish battalions were assigned to other parts of the Congo. But our destination was the capital, then called Léopoldville. Our base there was Camp Martini, previously a liquor distribution yard as the name suggests. It had a bungalow at the roadside. A railway line, serving the whole industrial estate of which it was a part, ran along at the back. In between were spacious sheds.

On the first night most of the officers were temporarily packed into the bungalow. Overnight there was an overflow of water from somewhere which 'miraculously' passed by the kitchenette, my bedroom, and flooded the kitchen, thus marooning the three or four officers whose camp beds were pitched there. As I emerged in the morning, I saw our legal officer, Pat Liddy, sitting up disconsolately in his bed, rubbing the red spots that were all over his face and neck from mosquito bites.

This first bad impression of humid Léopoldville never left us. As the last of the returning 36th Battalion passed through our camp, telling us of the beautiful climate, 5,000 feet above sea-level, in Katanga, our self-pity increased.

I have read that the Belgian government, in co-operation with the Catholic Church, provided the Congo with the best system of primary and secondary education available in Africa. However, they had provided no third level education. To make up for this omission, the Catholic University of Louvain had built a new university on a height outside Léopoldville and called it Louvanium. It had opened just a couple of years before we arrived.

Some of our troops were assigned to a post there. We were all glad to use its lovely swimming pool. Captain Browne, one of our staff officers, completed his accountancy exams there, being supervised by an Irish lady who was on the staff of the university.

After about six weeks, an anti-UN women's riot broke out in what was then Elisabethville but is now called Lubumbashi. General Seán McKeown (who before and after his Congo service was the Irish Army's Chief of Staff) was then UN commander in the Congo. He transferred us to Elisabethville at the request of the UN Indian commander there. We weren't sorry.

Léopold Farm was a former dairy farm with residences and outhouses, on the outskirts of Elisabethville. This became our new headquarters. The redbrick cowhouse occupied three sides of a quadrangle. The tiled roof sloped down to the inside of the quadrangle where it was supported by widely spaced pillars. One side of the cowhouse was converted into a chapel, the other a soldier's billet. It was here that Private Fallon was killed by flying shrapnel as he sat on his bed about a year previously.

In the surrounding area there were many, perhaps hundreds of bungalows, on jacaranda-lined avenues, which had been vacated by Belgians when fighting broke out. These became the homes of the officers and men of the battalion. I occupied a room in one of them which was also a medical centre for Dr Dan Sheehan. We were each given a local bicycle called a 'jambo bike'. Local people used to give us the Swahili greeting 'Jambo', meaning peace. A short distance away the senior chaplain, Fr Colman Whelan from Renmore Barracks in Galway, shared a house with the commanding officer, Lt Col Don O'Broin and a few other officers.

'B' Company lines were on the edge of open countryside of rich, red earth. Anthills were dotted all over the place like the cocks of hay we used to have before baling was introduced, except that they were much larger. One day, some soldiers went exploring around these anthills and were captured by hidden

20. Officers of the 37th Battalion meet President de Valera before their departure for the Congo in May 1962. Fr Ronald Neville is in the front row, left

Katangese soldiers. Fortunately, one of them escaped back to us and the rest were returned, after some negotiation, the next day.

So far as I can recall, that was the nearest we in the 37th Battalion got to a fight with the Katangese army. This was in contrast to our predecessors, the 36th Battalion (Private Fallon's) – who took part in the capture of Elisabethville for the central government – and to our successors, the 38th Battalion, who were actively involved in the complete abolition of the Katangan secession.

Fr Joe Fagan (who was Parish Priest of Dundrum when he died) whom I had temporarily replaced in Baldonnel aerodrome in the latter half of 1961, was chaplain with the 38th Battalion. In his previous tour of Congo duty in 1961, he had been captured along with a group of Irish soldiers and incarcerated in the mining town of Kolwezi. During the military push out of Elisabethville by the 38th Battalion, he later told me, he was hearing the confession of one of the soldiers when a bullet from a sniper whizzed by them both. Later, after they had captured Kolwezi, he had the satisfaction of setting up a chapel in the very room in which he had been incarcerated two years previously. That same room also became our daily Mass centre when I replaced him with the 39th Battalion in May 1963.

Lt Col Dempsey was the commanding officer. He had a good speaking knowledge of French which was very useful in dealing with his opposite number in charge of the Congolese national army battalion, also stationed in Kolwezi.

Fr Colm Mathews, later Parish Priest in Firhouse, was our senior chaplain.

We lived in the same bungalow and were allowed a Land Rover between us. We were part of HQ Company which was accommodated in bungalows, previously occupied by Belgians, on the edge of Kolwezi town.

'A', 'B' and 'C' Companies, under the command of Commandants Jack Ahern, Eamonn Young (a famous Cork footballer) and Bill Madden respectively, were billeted about two miles outside the town in a former army barracks. Commandant Jack Connole, our engineering officer, was appointed camp commandant to keep the peace between them! Beside the camp was a well equipped boys' technical boarding school, built by the Union Minière company and run by Belgian Salesian Fathers to supply technicians for the local mining industry. They allowed us to use their college chapel for our Sunday Masses.

The 'Bonne Auberge' was the name of the local hotel. I recall putting on a tie instead of a Roman collar for the first time since my ordination when the commanding officer took a group of us out for a meal there. It became a popular haunt for socialising UN personnel, which somewhat compensated its racist Belgian owner after the departure of so many Belgian patrons.

As regards a chaplain's activities, we followed the precedent set by Fr Paddy Crean, an ex-British army chaplain, who was chaplain to the first Irish battalion that went to the Congo in July 1960: daily morning Mass, evening Rosary, confessions and pastoral visitations to outposts. I personally interviewed as many soldiers as possible and wrote reassuring letters home to their relatives. I got some lovely letters in return but I have kept none of them.

The Church in the Congo was well staffed with priests but they were all Belgians. I just can't recall meeting any native Congolese priests. We always established good working relations with them. English speakers among them would willingly supply for confessions and even Masses. One such was Père Paschal, the Parish Priest of Kolwezi. He seemed to be popular with everybody. He had a beautiful parish church (built by Union Minière, of course) which he would kindly put at our disposal for special occasions. For such events, we would have a special choir, composed of all ranks, to sing the Missa de Angelis in Latin. All Masses were in Latin in those pre-Vatican II days. When I emerged from the church on one of those occasions – I think it was on 15 August – I saw Père Pascale and Lt Col Dempsey talking to General Mobutu, then commander of the Congolese National Army. From being a sergeant in pre-independence days, he was made supreme commander after independence.

Ordinary Sunday Masses in HQ Company had to be in the open air as the numbers attending would be so good. There would be virtually 100 per cent of the officers, and a very good attendance of all other ranks as well. Only on special occasions would there be a parade. I think we did have Sunday parades to Mass at the beginning of the tours of duty. By the time I went on my last overseas tour of duty, in 1978 to Southern Lebanon, times had changed greatly.

20

To the Congo with Some Rosary Beads

Fr Colum Swan worked as chaplain to the 2nd Infantry Group from November 1963 to May 1964. This is the story of his involvement in the UN peacekeeping mission to the Congo.

My commanding officer was Lt Col Redmond O'Sullivan. The unit headquarters, together with HQ company, was at Kolwezi in Katanga. An additional company was based at Camp Ruwe, adjacent to Kolwezi. In addition, an infantry platoon and cavalry element were in charge of Kolwezi airport, around five miles south of the town. A platoon was stationed at Lualaba Bridge on the road to Elisabethville, about twenty-five miles to the south.

21. Fr Colum Swan was an Army Chaplain
with the 2nd Infantry Group 1963–1964

As I was the only chaplain to the group, there was considerable travelling involved. Sometimes I drove myself; at other times a driver was made available to me. One experience stands out in my memory. It was probably the month of March 1964 and on a Saturday evening I was preparing for Sunday Mass. Making use of an army vehicle, with driver, I visited the platoon at Lualaba Bridge at about 7 p.m., an hour or so after nightfall. On the return journey to Kolwezi, perhaps two hours later, our vehicle cut out, dead stop, engine totally immobilised. We had no radio, no weapons, and the immediate danger was from roving ex-gendarmerie who roamed the bush – the 'carpetbagger' residue of the conflict – seeking whom and what they might devour. Furthermore, it was unlikely that we would be missed by HQ for two or three hours. What to do? We, the driver and myself, considered our situation. My recollection is that he was a Private Dunne from Portarlington. To wait at the vehicle was out of the question. To the north east, around one mile distant, there was a sound of tom-toms and we could see a reflection of a bonfire in the branches of coconut palms at that point. Obviously, there was an African village there, in celebration – probably of the Lunda tribe. We decided to head for the village. We had no idea as to how they might react to us, their celebrations might involve alcohol or hemp, and besides, we would have to walk through a mile of elephant grass five feet high to reach them. We failed to find any track towards the village and our immediate fear was of treading on a snake. We set off and said little. When we had covered about half the distance, I asked the soldier if he was fearful. He replied: 'I am, but isn't that what we are paid for?'

We reached the clearing of which the village of straw huts was the centre – from the grass to the huts being a clearance of thirty yards or so, all around. The village itself, comprising about thirty huts, was somewhat rectangular. In the centre was a blazing fire on which was the customary metal pot of enormous size. But what was sinister was the silence. In our anxieties in finding our way, we had not noticed the drums falling silent and now there was not a person to be seen. Our approach had been noticed and the whole village had withdrawn into the elephant grass. Though we saw not one, we were perfectly in their sight. The huts had entrances facing inwards. We walked along the west side. At every 'door' I called 'Jambo' (peace) and 'bonne nuit' (good night), but with the exception of one, where some poor old person cowered back in fright, there was no response.

At the first corner, I said to Private Dunne: 'We'll try the next line of huts [on the north side], then we must run for it, back into the grass.' Again, there was no response from any hut. At the last hut, I took a deep breath, probably blessed myself, and was about to say 'Now', when I heard from behind 'Soldat' (soldier). At least, this was better than the arrow or spear I was anticipating. Coming towards us out of the bush was a tall African followed by two or three others. 'Soldat, êtes-vous ONUsiens? Je suis le moniteur dans l'école ici. Est-ce que vous êtes officier?' (Soldier, are you from the UN? I am the local school instructor. Are you an officer?).

My inadequate French was a lifeline. I replied: 'Je suis un aumonier' (I am a chaplain). 'Je veux avoir un chapelet' (I want rosary beads) was his response, which luckily I provided and which he placed around his neck without delay – to him, a significant charm! With that, the whole village dashed out from the grass, each shouting 'Père, donnez-moi un chapelet' (Father, give me rosary beads). I explained to them that I had only one at that time but when, with their help, I again reached my camp, I would return to them with 'beaucoup de chapelets' (lots of rosary beads). We were now the best of friends and Private Dunne and myself were invited to join them in their supper from the big pot. With much expressed regret, I explained to them my urgent difficulty that would not permit of their gracious hospitality. The moniteur pondered the situation and then asserted: 'Nous allons a la gare pour téléphoner à votre commandant. Elle est a cinq kilometres d'ici' (We'll phone your commander from the railway station which is five kilometres away).

We reached the station, accompanied by the moniteur and his two friends, phoned HQ and transport was duly provided. In due course, I returned to the village and brought them the chapelets (rosary beads), which I trust they found effective in warding off evil.

21

The View From Léopoldville HQ

Vincent Savino was posted to UN headquarters in Léopoldville where
he served as a staff officer from mid-1962. He served again with the
UN forces in the Lebanon and, later in his career, rose to the post of
Quartermaster General of the Army, with the rank of Major-General
(1987–89). He is now president of the Irish UN Veterans' Association.
Here Major-General Savino looks back on the Congo years.

I was posted to the Congo in May 1962 as a staff officer in UN headquarters in the
military personnel section in Léopoldville, now Kinshasa. That section was
responsible for pay, allowances, leave, debts, funerals and policing. We were based
in a building called the Royale which had been a big hotel. The entire UN HQ was
more or less based there, including operations, intelligence, personnel and logis-
tics. The chief of staff was an Ethiopian general. The chief administrative officer
was there along with all the top civil servants. I flew out from Baldonnel with the
American transport service. I travelled with a group of staff officers and a platoon
of soldiers from the Western Command who were commanded by Lt Gerry
McMahon who later, as a General, became Chief of Staff. An interesting thing
happened on the flight which points up many of the problems we had over the
years. We were no sooner airborne than I noticed a soldier from the platoon walking up and down the aisle of the aircraft. He was going to the toilet at the rear of
the plane and back again. A few minutes later he would be up and down again. I
soon realised that there was a problem aboard. Evidently this guy was very gullible
and his fellow soldiers had been ribbing him and telling him that the plane was
going to crash. When we got to Tripoli in North Africa we disembarked. Whereas
at Baldonnel I had been seen off by all sorts of officers, including a Commandant
who carried my bag to the aircraft and a Captain who gave me my bag of sandwiches with an apple and orange, at Tripoli we were met by a US Air Force Lance-
Corporal who brought us to the mess hall while the plane was refuelling. When we
reassembled in readiness to board the aircraft again this particular soldier had gone
missing and we delayed the flight for some time until he was rounded up. He had
wandered off. When we got to Léopoldville that guy was immediately admitted to
hospital so he never actually served as a soldier. He suffered from some sort of
stress disorder. He had cracked up when the plane took off.

I have recollections of a similar incident happening when I was going to the
Lebanon to take command of a battalion there but it was more common in the

Congo days because soldiers were not even aware of where they were going. By 1962, though, some of them were going back for the second time. Many of those going to the Congo for the first time, however, didn't know the sort of conditions they would have to endure. I saw soldiers at a swimming pool we had access to in Léopoldville. They had big open sores on their arms where the smallpox inoculations had not healed up. Many of the guys were sick and there was always a fair number of Irish soldiers in the hospital in Léopoldville. I counted nine there at one stage for various problems, most of them psychological. It just shows you that in those days soldiers were so unprepared for what hit them in the Congo. The heat and humidity were very trying. I was looking forward to having a bath of cold water but the water wasn't cold. As soon as I got out, dried myself and got into clean clothes I immediately began to sweat again, so I was back to where I was at first. When I went out there we had light-weight uniforms but they were still heavy enough.

The department I worked in was very cosmopolitan. My boss was a Malayan Lt Colonel. The second in command of the section was a Canadian major. The staff officers included Captains from India, Pakistan and Ethiopia, as well as myself from Ireland. There was a huge orderly room staff that had to undertake all the documentation for the UN forces arriving and leaving. Within that large office there was a Pakistani jemadar [junior officer], who was the chief clerk, two Irish sergeants, two Malaysian corporals, an Indian Sikh sergeant, several Congolese clerks, a Swedish secretary and several Norwegians.

I went to Elisabethville on a mission, as well as to Kamina and Luluabourg. At that time Katanga was reasonably quiet. The climate in Elisabethville was absolutely different because it was East African, drier and cooler, so there weren't the same problems we had in West Africa. The Congo was a huge place. The trip from Léopoldville to Elisabethville by air took five hours, which is the same time it now takes to cross the Atlantic. We had a lot of health problems in the west coast region, including skin disorders from sweating. It was hard to sleep at night.

I met Belgians there although there were not many of them. They were very tentative about the situation. They were not very sure how they were going to survive. Several of them were business people. There was a brewery which produced quite good beer, and that was run by Belgians. They continued to run it after independence. The restaurants in downtown Léopoldville, and even the pubs, were run by Belgians. I vividly remember going to a pub there on a Saturday afternoon with some of my colleagues and seeing the results of European cup football matches on the notice board, including Shelbourne. It was a great joy to see that. Communications with home were very bad, though. When I was out there my father-in-law died suddenly and I could hardly communicate with home. The radio link was very bad; most of the communications were by telephone. I managed to get a phone call through eventually but unfortunately there was nobody home. They were the difficulties we faced. When people got sick and went to hospital or died it took a long time to communicate with home.

I met several mercenaries who were hanging around. Most of them weren't impressive, they were just hangers on who were in it to make easy money while they could. They were well paid but as soon as any trouble started they ski-dadelled. We had a Congolese army liaison officer working with us and through him we had good contacts with the Congolese army. We actually played a football match against them. A lot of the UN people were high calibre footballers, including several internationals. One of them was a Colombian, another was a Ugandan and another was from Ghana. They all played on our team but, despite this, the Congolese hammered us six goals to one. I scored our goal. We had an inter-UN league in Léopoldville. The Argentinians had an air force contingent and there was also a Danish military police company. There was also a Nigerian football team. We tried to maintain a normal lifestyle there but it was very difficult because there were always shootings downtown or people otherwise getting into trouble. By and large though, it was a very interesting experience for me and I enjoyed it thoroughly, I must say.

There was a civil war going on at the same time. In the operations section where I worked there were always messages coming and going about incidents throughout the Congo, but we seemed very remote from them. The distances were such that very little could be done. Originally, when messages came through that there was fighting or a massacre somewhere you wondered what you could so. Once the operations people came on board they could do very little apart from sending a reconnaissance aircraft to have a look the following day. The reaction time was very slow and naturally enough we did not have enough troops to police the entire Congo.

Quite obviously, Katanga was the part of the country that had the riches, including mines which were very important for the well-being of the new state. If they had let that go, the rest of the Congo would have been bereft of sources of revenue. The decision not to allow the Katangese to secede was a good one but it cost lives, of course. I do not think the ending of the secession could have been negotiated because the people running Katanga were only the front people for Belgian financiers. I imagine that the decision to go in and stop them from seceding was the correct one. However, it cost a lot of lives.

One has to go back and blame the colonisers who, at the beginning, never prepared the people for independence and who left them with no infrastructure whatsoever. For instance, Mobutu, who had been a non-commissioned officer in the Belgian army, was suddenly thrown in as the top man. The whole culture in Africa, West Africa in particular, is one of making what you can while you can, and getting out. We saw this in all the other countries also. The other thing the Belgians did – which was entirely wrong, to my mind – was to make a very quick exit from the Congo. They just declared independence for the Congo and got out without making any attempt to hand over power to whatever responsible people might have been there. The result was that thousands of people lost their lives and it continues to this day.

I would prefer to have seen a different sort of Congo evolving in 1960.

Whatever they say about the French, at least they left a better infrastructure behind in their former colonies. They also educated a fair number of the local people. In addition, the French maintained contact and a military presence by treaty in many of their former possessions. For instance, I remember going across the Congo River to visit Brazzaville, the capital of the People's Republic of the Congo [a former French colony]. There was a huge difference between it and Léopoldville. Brazzaville was a bright modern city with all sorts of goods available in the shops. We used to go shopping there; it was a bit of an event. It was well run at that time. There were French paratroopers there, as well as French Foreign Legion troops, in evidence on the streets and everything was running beautifully. But it, too, collapsed in due course.

As far as the Irish Army was concerned, the Congo was the best thing that ever happened. At that time the army was really in the doldrums. There were only about 7,500 men, many of whom were not capable of going overseas. Suddenly we had 1,200 men going to the Congo with the 32nd and 33rd battalions, with only three months between them. The Army was rejuvenated receiving new equipment and great experience. There is no doubt that we were very good peacekeepers. It was the making of the modern Irish Army. In 1964 when we finished in the Congo, almost immediately we went into Cyprus with the United Nations. From Cyprus we went on to the Sinai and afterwards, in the 1970s, the Lebanon came up. There was a continuation of overseas service and the Army became very professional. In 1969 when the troubles in the North started, we were better able to cope with border duty than we would otherwise have been. At first, that was disastrous but we coped because we had people who were competent and had gained great experience overseas.

In the Congo and in Cyprus we proved that we could take over at the top level and produce people who could successfully command multi-national forces. The Irish Army had people working at all levels in the UN. Several senior Irish officers were working where I was based and they were all fair minded and highly thought of. They did not become politically involved and they did a great job. That tradition has lived on to today where we have people serving overseas all over the world. At the moment we have people in New York, in Vienna, in Bosnia, Kosovo, Kuwait and in the Middle East. The latter has been a very successful operation and is the nucleus for all UN operations in the Middle East and the East generally. We have done well and have improved ourselves. Now things have moved on and we have to struggle along and keep up that great tradition.

We lost many people, of course, and the memorial outside Arbour House in Dublin bears the names of 77 soldiers who died on UN duty. That figure does not include the number of people who came back scarred, either physically with wounds or others who were psychologically damaged. That was the biggest problem, particularly for the older guys in the Congo. They went out unprepared psychologically, in particular. At the time of the Congo, soldiers were largely uneducated. It was the tradition then that young men leaving orphanages at 15 or 16 years of age went straight into the Army. Some guys were wounded while fighting in the Congo at only 15 or 16 years of age. Such people served until they were 60 years old and faded away after that. Nowadays it's a lot different. Young men coming into the Army are educated and they are much more sophisticated, but in those days they weren't.

There is a great story about a guy from the Western Command who went home to tell his wife he was going to the Congo at short notice. She wasn't at home so he wrote on the door: 'Gone to Congo'. His wife thought he had gone to Cong in County Mayo and that he would be back the following day. He didn't turn up for six months. That sort of thing happened. People just didn't know why or where they were going or what was happening. If you asked a lot of them about the geography of Africa they didn't know. People did not travel as much in those days. Young men now are well travelled. Many of them have visited Europe or America before joining the Army so they are much better able to cope. It is a different world and the communications are so much better. They bring their mobile phones with them and can call home at any time. They can come home on leave half way through their tours of duty, whereas nobody got home from the Congo before their tour of duty was up, except in a wooden box.

22. General Seán McKeown (centre), UN Commander in the Congo, presides at a briefing of UN Officers, Léopoldville, 1960

22

Memories of an Irish UN Liaison Officer

*Dubliner Pádraig Ó Siochrú served as a liaison officer with the
United Nations peacekeeping mission's headquarters in Léopoldville.
Here he remembers the highs and lows of his time there, towards the
end of the mission in 1963–64.*

I felt I wanted to be in the army. I had a cousin in the army and I said I didn't
want to go into the bank or anything, so I decided to join the army, just like that.
I had seen enough civil servants around me and I was not interested in the financial sector either. I did twenty years in the army. I was based with the 1st
Battalion of the Western Command in Galway – the Irish-speaking battalion.
Later on I was with the FCA in Belmullet, County Mayo. I was called back to
the Military College to do translation work when the Minister for Defence,
Kevin Boland, decided to Irishise the army. After leaving the Military College
in 1966, I joined RTE's presentation department with Pádraig Ó'Raghallaigh.

Someone said that a hard neck and an ailing wife was a good way to get to
stay where you were in Dublin. But I found people who would not come down
to the Curragh, yet they were able to go to the Congo. I told one colonel, 'Isn't
it gas? There's two of them there who could not come down here from Dublin,
yet they are out in the Congo at the moment. And here am I – I've been here
for years with you.' He said, 'Okay. I'll see that you get your turn too.' So, I
was posted as an Irish liaison officer to the UN headquarters in Léopoldville.
My job was to liaise between the UN and army HQ in Dublin. I also liaised with
the 3rd armoured cavalry unit in Léopoldville and the 36th Battalion in Kolwezi
in Katanga. I saw messages go through. It is hard to define exactly what the job
entailed but if there were complaints, someone visiting, a bit of trouble or somebody missing, you had to be delicate about it.

I studied a little bit of French there but not enough. I remember a Belgian
professor asking us all if we could speak French. There were about ten of us:
one Englishman, myself, and the rest were Indians, Pakistanis, Norwegians and
Austrians. Then he asked us all to speak in our own language, so I blasted away
in Irish. He said to me: 'Ah, mon capitaine. Vous avez les sens de français!'
(Ah, captain. You have a good grasp of French).

In November 1963, I remember drinking with members of the American
embassy in Léopoldville when it was announced that Kennedy had been assassinated. A gendarme came into the hotel to tell the female diplomat: 'Excusez-moi,

23. Léopoldville airport, late 1963, left to right: Judge Advocate, Colonel Art Cullen; Cmdt Chris Dawson; UN Liaison Officer, Captain Pádraig Ó Siochrú – on their way to a court martial in Elisabethville

Madame. Votre President est mort.' That's how I heard of the death of President Kennedy, while drinking a bloody mary. We were all terribly shocked and couldn't believe it. It is so long ago, but I think we just had another drink and stayed silent, looking at each other. We had got to know members of the American embassy staff. They were very decent chaps. All the embassy staff, and the American military air transport service team, spoke French. They had all been told to do a crash course in French for thirteen weeks before going to the Congo. Kennedy was taking the thing seriously up to his death in November 1963.

I was there from August 1963 to February 1964 when the UN operation was being wound up. As a matter of fact, we got on so well with the man in charge, General Ironsi of Nigeria, that he wanted all the Irish – the whole unit – to come back to Lagos with him in a boat, and stay there.

On one occasion the Ethiopian troops came up from Elisabethville to Léopoldville, and apparently a few of them had a jar. They didn't kill anybody or hurt anybody but the Emperor, Haile Selassie, heard about it and sent an order to the battalion commander to take out every tenth man and shoot him. The commander said it couldn't be done, but the emperor was adamant altogether. Ralph Bunche, who was Assistant Secretary-General of the UN, left New York for Addis Ababa to persuade Haile Selassie that this could not possibly be done. The emperor accepted that but he sent the battalion commander a message, which stated: 'If anything like this happens again, you'll lose your head when you come home.' I thought that was beautiful – 'you'll lose your head when you come home.'

On another occasion, I remember meeting an American Air Force major in Léopoldville, whose language was pretty strong. He said he'd got seventy-two hours' notice to deliver a plane. This was what you would call an intelligence or spy plane. It was a present from President Kennedy to General Mobutu. The air force major was effing and blinding because he only had seventy-two hours to bring the 'bloody' plane out to the Congo. The aircraft was a real revelation – a present from America to the Congo, just a few weeks before Kennedy was assassinated. The Americans were clearly backing Mobutu.

At the time, Seán Ó Riada's 'Mise Éire' had just come out and General Seán McKeown, who was the UN commander there, had brought three or four copies of the record out with him. When we finished our work, we used to play the music every night. I brought six copies of the record out with me because I thought they might make nice presents. There was a party one night and some-body said, 'Pádraig, did you bring that record with you?' I had it with me and so it was put on a record-player and the next thing, there was dead silence. An Indonesian lieutenant colonel grabbed a chair and stared at the gramophone. When it was over, he said: 'Captain Ó Siochrú, that is great patriotic music.' So I gave Mise Éire as presents to people like him and they were thrilled with it. It ended up in all sorts of places, including Indonesia.

I had to travel to Léopoldville airport quite a lot and I was told always to keep a revolver in the car, just in case. I didn't have to use it and nobody shot at me, but an Italian air-force captain told me: 'Listen, this may sound terrible to you, but if by any chance you knock anybody down going to or coming from the airport, don't stop, because if you do they will kill you.' He explained that was because the locals believed they had to avenge the spirit of the person who died by killing the person responsible for their death. Had it happened, I would have been hacked asunder. It was an unusual thing for the Italian to say and I thought he was being cruel but he wasn't – it was a warning to me. Thank God, nothing like that happened but if I had not been told about it, I might have stopped and then been killed.

On one occasion, I travelled down to Elisabethville or Kolwezi for a court-martial with the judge advocate, Colonel Art Cullen, and Comdt Chris Dawson. We flew out in a 30-year-old Dakota DC-3, which had been stripped down of everything, so we were sitting on mailbags. The Israelis were maintaining the aircraft and the pilots were Brazilian. They said to us: 'Will you please go for a walk gentlemen, because we can't start the second engine yet.' So Cullen said to me: 'Pádraig, do you know your Act of Contrition?', and I replied, 'I do sir, in Irish'. But we landed safely anyway, after an 800-mile flight during which we saw giraffes and lions. If the second engine had gone, you would never have heard of us again because there was 800 miles of sheer jungle and desert. The court-martial involved an Irish soldier but I have forgotten what the charge was. He got off with a fine in the end.

General Ironsi was the first commander there. He was a fine character from Nigeria. He was six foot, three inches tall. We got on terrifically well with him. When he returned home to Nigeria he became Prime Minister of the country and was later assassinated. I always remember that he had a great sense of humour. The

UN headquarters in New York decided that liaison officers weren't necessary and could be discarded. They sent him a long twenty-page memorandum outlining the reasons for this, but the general took out a biro and across the top of the UN memo he wrote in capital letters 'B-A-L-L-S'. That was what Ironsi was like – he was lovely, a terrific character in that way. He was very direct. He had been trained by the British who built up the native men, appointing them to officer rank, including lieutenant colonels. The Nigerians were very helpful to me in the Congo.

There were some Belgians in Léopoldville in the 1963–64 period, but most of them had left. It was a beautiful city and when the Belgians ran it they used to hose it from the air every week to keep it clean. When they left, everything was dropped and the place was in a bit of a shambles, including the airport. Nothing much was done but the Congolese were only starting off on their own. At the time, Mobutu was a lieutenant general and I met him at a church meeting. I have a Christmas card he sent to me. I remember getting an invitation to attend the Congolese national day, which is 17 November. They had a Mass, a march past and an air display. I remember the air display: out came one parachutist flying the flag of the Congo; a second parachute carried the battalion commander; and the third parachute carried Lt-General Mobutu. He hit the ground first, threw off his parachute harness and sat in a regal chair to address the troops in French, not Lingala or Swahili. Kasavubu was the president. He used to go around everywhere in a stretch-limo, with hooters going. It was real Hollywood stuff. He was a very small man but Mobutu was bigger and was respected in the army.

In the light of current events, including butchery, killing and humiliations that are going on in Iraq, and chopping off heads, the Irish UN troops in the Congo proved how disciplined they were. A few days after the Niemba ambush on 8 November 1960, Irish troops saw rifles being taken off the Balubas. They could have retaliated then, but nobody fired a shot. You don't do that revenge thing. It was a question of discipline – retaliation was not allowed and would not be done. It would have been revenge and would have made us look bad, because we were peacekeeping troops. Unfortunately, Lt Kevin Gleeson, who led the Niemba patrol that day, did not have enough ammunition. Others might have fired at the Balubas and it would have been totally justified, but there is no doubt that Gleeson was too nice. But that is the way it was – nobody did any shooting.

When I was in Léopoldville, the blue berets were accepted by the Congolese and there was no animosity whatsoever. There was very little fighting, except at Jadotville, which has been mentioned in the news recently, and rightly so. When Jack Quinlan, the man who was in charge at Jadotville, brought the troops back, he was welcomed in Athlone and cheered all over the place. His job was to bring his soldiers back, it wasn't to fight a battle and kill. The truth is coming out now, that the failure to recognise their contribution was unfair. Some veterans of Jadotville, including Liam Donnelly, say they don't want any awards – just that the record would be put straight.

There wasn't much trouble in Léopoldville. The cavalry was out there and it was a question of keeping the peace. Things had worked well there but, God, you see the way the Congo turned out eventually. Should the UN have stayed a bit longer after 1964? I would say not, given the way things were. Law and order took over. General Ironsi was the last force commander there. Nobody fired a shot in Léopoldville for the whole period, while I was there. Very little else happened with most of the other battalions, bar the Jadotville incident and the Balubas at Niemba. It was a peacekeeping mission and the peace was kept. So the UN felt they had to go elsewhere, off to the Lebanon and Cyprus.

The Irish language was used a fair bit, particularly to pass secret messages. They thought no one outside our army could speak Irish in the Congo, but I was told there was a native speaker from Connemara who was working with the Rhodesian police force. It was said that he knew every word they were saying over the radio, and that he let them know it, too. I cannot vouch for the veracity of that story, but that was the *scéal* at the time. The Swedish all spoke Swedish, English and French, at least. I remember fellows who didn't speak Irish to me here, but they spoke it out there in the Congo – not because I was speaking it to them, but because they wanted to show other UN contingents that we also have our own language. When they heard Indians, Norwegians, Finns and Swedes, not being rude, but speaking to each other in their native languages, that got the Gaeilge out quickly, honest to goodness. I remember that an Irish-speaking Jesuit priest, Fr Colm Ó Ríordáin, came in from Lusaka. I was so thrilled at the prospect of being able to talk Irish to him, that I drove along the wrong side of the street in Léopoldville on my way to meet him. I was driving on the left-hand side of the road, like at home, instead of on the right, but I didn't hit anybody.

We got on all right with the Swedes. Their work was God, or God was their work. We found that you had to explain jokes to them because they hadn't a great sense of humour. The Norwegians, on the other hand, were very like us in temperament. The Swedes were very serious, very good workers and tremendously able. The Swedes were all doing their military service in the Congo and quite a few were university students. The Swedish army had a brigade ready to move to places on UN service. It was all seriously sewn up and ready to go.

There was a variety of nationalities at the UN peacekeeping mission's headquarters in Léopoldville. The advisor to the force commander was an Irish engineering corps officer. The UN had a very good technique. They always put a Pakistani and an Indian working together, with the purpose of making them cooperate. We got on very well both with the Pakistanis and the Indians. On one occasion, I remember an Irish officer colleague who died out there while sleeping in the afternoon. When I was told he was dead I remember reciting the Act of Contrition in his ear, in Irish. A Pakistani commandant came to me later and said: 'Captain Ó Síochrú, we have cancelled all entertainment for tonight in honour of your comrade.' The Pakistanis didn't drink, of course. I used to love the meals they'd have: they always served jugs of water with their hot, spicy curries. I loved the curries but, janey mack, you couldn't eat them every day – God no. We got on extremely well with the Pakistanis. I remember the sound of their voices – you'd swear it was a Cork man talking to you.

<h1>23</h1>

<h1>The Foot Soldiers' Stories</h1>

Regular troops who served with the UN peacekeeping mission to the newly independent Congo, from 1960 to 1964, recount their many and varied experiences of army life in Africa.

24. Joe Desmond

JOE DESMOND, DUBLIN

My first trip to the Congo was a very enjoyable one, although one lad lost his legs. I was with the 34th Battalion. We relieved the 32nd and 33rd Battalions in Kamina. I was there with my battalion when we moved into Elisabethville. After returning home to Ireland we went on a second tour of duty with the 36th Battalion. That trip was not very pleasant. I was only 18 years of age. As we flew into Elisabethville airport one of our lads, Mick Fallon, was already dead, lying in a coffin. I trained with him in Dublin. He was killed by a mortar shell. He was attending Mass when the shell came in through the church window. He was killed outright, while a few of the other lads were injured.

We went into a war situation. I was the number one mortar man and at the time Sgt Paddy Mulcahy had been wounded on the railway tracks. He was left in the rear and later died on the railway tracks. We were moved up to a place called the Liège crossroads. That's where we were firing from. I was put away from the rest of the lads because I had a loud voice when I was giving orders.

So, I actually got the orders concerning what positions to fire at from the observer out in the field, and I gave them back to the mortar crew. All of a sudden the shells started landing all around us. One landed on me and I was cut to bits. I was wounded in the chest, the arm and my leg. I lost a lung and spent an awful lot of time in hospital. Sgt Mulcahy actually died beside me. Lt O'Riordan was already dead and Andy Wickham had been killed. We came under fire from the gendarmerie who had a lot of mortars. The mortar fire was coming from them and from Tshombe's army. There were a lot of mercenaries too.

The Irish – A-Company – actually took the tunnel [a railway bridge over a road at the entrance to Elisabethville] but many troops of other nationalities also lost their lives. It was the most highly decorated company in the history of the Irish Army. We had mortars also. They had pinpointed our position and although we had pinpointed a lot of theirs, we did not know the result of our firing. You never know the results, you just fire. We made direct hits, we may have killed a lot of people, but I don't know. You don't want to think about that, you don't want to know. There were five killed with us on that trip and a good few were seriously injured. I spent a lot of time in hospital myself afterwards. I had a big operation in the Congo, there and then. There was a little garage outside where they had put all the coffins. They brought me out one day for air and I could see all the coffins. The hospital was like an ordinary house. Fellows were not laughing for too long. All nationalities were brought in to that house which had been commandeered as a hospital. Fellows were dying, so they just put them into coffins and took them away, you know.

I spent time in Elisabethville and later at a hospital in Léopoldville. I was also in hospital in Italy, at Camp Darby, a United States airbase. I also spent time in a hospital in Germany, at an American airbase there. I spent a long time there. I came home after six or seven months. I was the only person on the plane when it landed at Dublin airport. I was brought to St. Bricin's Hospital and the next day I was removed to the Richmond Hospital. My chances of recovering were very slim, but I did recover.

The fighting for control of the tunnel in Elisabethville finished before Christmas 1962. It went on for a few weeks. It was the first time I had had a misfire with the mortar. We had gone through the procedure of taking the shell from the barrel, but it went off and burned my hands. My hands were all wrapped up. It was reported in the papers that one member of the company had been slightly wounded. Afterwards there were three or four who were seriously wounded, but the newspapers did not take up on that. People were too quiet to kick up a fuss about anything, so it was just left like that.

I think the Irish involvement in the Congo achieved something. The nation was very proud that their troops went overseas. We were there to make peace, not war. Since then we have become famous for peacekeeping all over the world. Irish peacekeeping troops always get on with the local people, just as we did in the Congo. The Congo was the start.

To this day, I still have some shrapnel in my chest. When I go to Lourdes every year, the airport security alarm goes off when I go through it. I have to tell them beforehand that there is a bit of shrapnel there. The scar from the wound goes from my arm right around my chest to my back. I consider myself very lucky to be alive. They actually thought I was gone. My parents were told that I was dead. I don't really remember the reunion with my parents when I came back home because I was too sick. They actually passed me by in the bed. I was very thin because I had lost so much weight. I could not talk because of a hole in my chest, and all the bottles and wires going into me. I was gasping for breath. My parents could not sit on the bed because the least bit of pressure would cause the pain to wrack through me.

Professor C.K. Burns operated on me in the Richmond Hospital, and I survived, luckily enough. They told my parents that I only had a slim chance of coming through. They did not think I'd make it but I did. It took me a long time to get up and walking again. I was in the Richmond for a long time. When I left the hospital, for a long time I would not go out on the street because I was ashamed. If I had to go out I would put my head down because I was ashamed after being shot. I felt everyone was looking at me saying, 'Look at him'. I felt I should not have been shot. That's the way I thought at that time. I did not know any better. It is only after, when you get older, that you realise it doesn't matter where you are, a shell can hit anybody in any position. So it didn't make any difference. I was young then, though, and I didn't understand. You could put it down to nerves. My nerves went for a while and I had some electric shock treatment in St Vincent's. I was okay after that. I left the Army in 1963.

25. Billy Lawlor

BILLY LAWLOR, DUBLIN

I am originally from Castlecomber, Co Kilkenny. I joined the Army in 1959 and served with the artillery. I went to the Congo in 1960 with the 33rd Battalion, A-Company. Our commanding officers were Col Bunworth, Louis Hogan and

P. D. Hogan. We were in Albertville when we got the word of the ambush at Niemba. At first we were told that it was a car accident, so we weren't really ready for what we were to face after. The company was formed up and we went out to Niemba. I met a friend of mine who I had joined the Army with. He was crying and I asked what was wrong. He said that a section of the company had gone out on patrol and had never returned. He said: 'I think they were ambushed.' We went in and had some food. The next thing, we were formed up, got into the trucks and went out searching for them. This was the day after the ambush. They had been out overnight and were reported missing at that stage. We went out through the bush searching near the bridge. Somebody shouted that one of the lads was staggering up along the dirt track. It was Private Kenny and he had two arrows stuck in him. We got him into the back of the truck and some of the lads brought him back to Niemba. They searched around and found the other bodies. The bodies were brought back in trucks to the village of Niemba. We had to identify them, wrap them up again and put them on the helicopters back to Albertville.

At the scene of the massacre, I saw Mattie Farrell who had been killed. We had to wrap him up. He was the only one I saw when the bodies came back. Most of them had arrows in them. Mattie Farrell had an arrow right through him and he was puffed up with the heat of the sun. He was unrecognisable really, you know. We believe they were poisoned arrows. Nine were killed. Tom Kenny and Joe Fitzpatrick survived. Joe was uninjured but he was very traumatised. Kenny had arrows in the back of his head and one in the cheek of his backside. He was in hospital for a long time out there. He was treated in Albertville where we were based.

I think Niemba was an unfortunate accident. A lot of people say it should not have happened, but we went on patrols out there and you met Balubas and other tribesmen. You just talked to them, with their greetings, and that was it. You kind of got used to it. It was unexpected when they did open fire. We were taken by surprise really. The Balubas used to create roadblocks and they dug up bridges. If lads wanted to do patrols they had to fix the bridges to get across them. It is as simple as that. I do not really know if that could have upset the Balubas. There were rumours at one time that a shot was fired at Niemba and one of the tribesmen was injured. They said that was what caused the Niemba ambush. That could be it – either the bridge or one of their own lads getting injured.

I nearly came under fire myself in Niemba. We were on a hill and the Balubas were around us, but there was no firing. There were stories of atrocities having been carried out among the locals. We went into one village and all the men folk were all missing. The girls said the men folk had been taken off by the tribes. We did not find any evidence of that or any bodies. My friend Teddy Brennan from Kilkenny did, though. When he went into Niemba they had to burn bodies which had been left lying around. I was with another platoon at that stage.

I think the Irish involvement in the Congo achieved something. If we had not been there there would have been slaughter all around. As it was, they were killing their own people and raiding. It was like the films about the Zulus. It was a similar story – attacking and killing one another. The 32nd Battalion came to Niemba to give us a hand and there were Pakistani UN troops there also.

We were very shaken up by what we had seen at Niemba. When I came back home I went on the beer. I had trouble getting a night's sleep. The missus told me that I used to have nightmares in my sleep. It played on my mind and it took a long time to get over something like that. There was no counselling provided by the Army. You went there, came back and carried on as normal. Nowadays armies do provide counselling but not then. Lads who go out to serve in the Lebanon now are facing a more modern type of warfare. In the Congo we were facing knives and bows and arrows. You didn't give it much thought, but if it was somebody with a machinegun you would be thinking twice. We were well armed ourselves. We had the old .303 rifles and I was a bren gunner. I never had to fire at anybody. We just mounted it up to be on the ready all the time at airports and places like that. The .303 rifle had a short bayonet on it.

26. Michael Butler

MICHAEL BUTLER, DUBLIN

I served in the 34th, 36th and 38th Battalions in the Congo. I did not experience major problems with my first and last tours of duty in the Congo, but the 36th Battalion was involved in the famous battle of the tunnel in Elisabethville. I was the sergeant in charge of mortars there. It was the biggest awakening I ever got because when I went there I thought I was an expert in mortars. I found out after ten minutes that I wasn't. We used to train at the Glen of Imaal where everyone knows the range, but it was very different in the Congo. The compasses were obsolete. They were made in 1915. You were trying to use those in a different area.

On the morning of 16 December 1961, I was in charge of the mortar platoon at Elisabethville. The tunnel was really a railway bridge running over a dual carriageway and a cycle path. There were train carriages on the railway from where fire was coming. A couple of our fellows lost their lives. I was supposed to fire a barrage of mortar bombs onto that target at the start. We should have had six guns but we only had three. I dug a hole, as I had been trained to do, and put three mortars into it. I soon found out, however, that it was a bad mistake. I was to fire the first bomb at 4 o'clock in the morning and when I did it was like flushing a toilet – the rain came down and I never saw anything like it before or since. As a result the mortar base plates turned over. We had no shovels so the mortar crew had to dig them out with their bare hands. We got one mortar firing.

The thing that struck me most was that we were firing over our own troops for the first time. I had no experience of it and I was lucky that I did not hit anybody. I was trying to follow the fire plan and for once the radio communications worked, which normally they didn't. However, that morning they did luckily enough. I still worry that we could have killed someone when I fired the last three bombs, once I had got the guns operational again. I found out later that some of our own troops had moved into a target area. We missed them anyway.

There were casualties on the other side. Firing the mortars, I was a couple of thousand yards behind, but there were casualties. They found about half a dozen of them, including one mercenary. I did not see them because when I got up there they had been buried. We were held there in readiness for a couple of days and did not go up until Christmas morning. That is when we heard all this. The fighting lasted a couple of hours. It started around 4 a.m., but I do not know when the actual attack went in. Some of the battalion were actually still in Dublin, so we went in under strength. From the radio conversations I heard, the Swedish went in and got slightly bogged down. We were lucky only to have had a couple of fatalities.

We were fighting against Tshombe's army with mercenary officers. Tshombe had his own officers but they really took their orders from the mercenaries. When I was serving with the 34th Battalion in Elisabethville, I spoke to one or two of the mercenaries and their attitude was that they would take Tshombe's money but they were not going to die for him. I met a couple of Scottish mercenaries and that was their attitude. They might have been privates in some army but now they were captains or majors. Some were even colonels. I am not going to mention any names but I actually met one mercenary from Cork and he had served in the Irish Army. I never saw Mike Hoare, a mercenary who, I am told, was from Rush, Co Dublin.

When I was there with the 34th Battalion we were invited out to an old place called Sabena Villas. We were in there with this mercenary – I think he was Irish, from Cork – who was sitting down and he had twin pistols. The waiters did not come quickly enough so he pulled out the pistol and blew out one of the light bulbs. He got attention then. We got drunk. I don't know who paid for it. We didn't and I didn't see your man paying for it either. That was a funny incident.

I was not surprised to find an Irish mercenary there. I toyed with the idea of doing it myself, at one stage. The Irish mercenary told me that he had served with the Irish Army during the Emergency. He was having a good time there, anyway. I was told that American dollars and English pounds were being paid into Barclay's Bank for the mercenaries but whether that happened or not I don't know. When Katanga seceded from the Congo, Tshombe printed his own money but it wasn't accepted anywhere else. The mercenaries were getting a living allowance in Tshombe's own Congolese francs.

I think the Irish involvement in the Congo did achieve something. When I went out there I had not been properly trained. I was a mortar man and a machine gunner but I had never been allowed to fire one of those weapons on my own. There were always two or three people breathing over my shoulder. At the time, because of the defence budget, you would fire perhaps three mortar bombs a year. In the Congo, at one stage, I was firing three bombs a minute with no supervision and a young untrained crew. I do not think the training was adequate. People may say that I did not apply the training correctly; I think I applied it in the way I was taught but it did not suit the terrain. The Irish Army supplied us with old No. 4 rifles but with the 34th Battalion we had Gustav machine guns. The Congo campaign was responsible for upgrading the army an awful lot. We learned an awful lot, I know I did.

History says that Mobutu, who took over from Tshombe, was a dictator too. However, in my opinion, the black men were used to taking orders or instructions from white men. They found it hard to take orders from themselves. Mobutu was an ex-sergeant major, as far as I know. He lasted a good few years but it will always be like that. It is tribal just like it is happening in other parts of the world; one tribe will wipe out the other and someone comes up, as happened in Uganda. That's Africa.

While it would be difficult to train people properly for the Congo terrain here in Ireland, we were trained from British Second World War manuals, in the old-fashioned way. It was for all out war, where there would be a line on a map, which related to a line on the ground. When anyone moved on the other side of it, you shot him. That was the old First and Second World War way of doing it. We went there as peacekeepers and you had to give the other first shot; you couldn't fire first. Even then, if someone was hurt, you might not be able to prove that he fired first, so that was a disadvantage. In addition to that, some of them wore uniforms while others didn't, so it was very hard to distinguish who was the enemy and who wasn't. I did three tours of duty there but I was only in combat for one or two days, although we were on red alert for sixteen days.

27. Michael Colton

MICHAEL COLTON, DUBLIN

I served with A-Company, 33rd Battalion. My most vivid memories are of when 180 of us arrived on a Saturday. We were stationed in a place called Kamina. Like the good Catholic soldiers that we were at the time, we were marched to Mass the next morning. The first thing we noticed when we got to the church was that every man and woman civilian was armed to the teeth both with hand-guns and rifles. We, however, hadn't got a rifle between the 180 of us. It brought out exactly what was happening out there when you saw women armed with handguns in a church.

Another vivid, and sorrowful, memory was the ambush at Niemba. It happened on 8 November 1960. There were eleven people on the patrol, of which nine were killed. The day after, a man named Davis was shot accidentally. I knew every one of them personally. I had been with Corporal Peter Kelly about two days beforehand in a place called FILTISAF [a cotton mill], a couple of miles from Albertville. We were great friends, and had been in the Army together in the 5th Battalion in Dublin. When I was out there, shortly before the ambush, one of my children died. At that time it was impossible to get home. Because of the death of my child, I was picked to come home with the bodies and I travelled home with them. It was the most harrowing journey I ever went on. It was absolutely dreadful. There was a KLM DC-3 aircraft with an American crew of five. There were six of us with nine coffins, if you could call them coffins. They were crates really. The bodies were stitched in blankets first of all and then put in ordinary coffins. They were then sealed into a tin coffin which was put into a sealed crate which had to be lifted by crane onto the air-craft. Each crate weighed 1,200 lbs. I stayed at home until 1 December 1960 when I headed back for duty again in the Congo.

I was at the scene of the Niemba massacre. The people died in different places. It mostly took place at the bridge. Their task was to keep this bridge open – it was only a plank bridge across a river – so that supplies could get through to Albertville. Most times when the patrols went out this bridge was down, and they re-erected it with the help of the Balubas. But on this occasion the Balubas had turned savage. Some of them were killed on the road immediately. Some of them made it into the bush. Others actually made it into a village and were killed there.

No one knows whether that incident stemmed from a Baluba having been shot earlier. Two months earlier a Baluba chief's son was shot in the head. What happened was that the people guarding the camp – No. 2 platoon of A-Company – had instructions to identify anyone who was entering it. This particular lad was entering the camp. He had a bicycle with him. A man named Davis stopped him. He threw the bike at Davis and ran around behind the back of the houses. When he came out the other side, Davis took a shot at him and hit him in the head. It turned out that he was a Baluba chief's son and it was shortly after he was released from the hospital that the ambush took place. But no one ever connected it with the shooting of the man. These things happened all the time when the Belgians were there. They just flared up at certain times. Personally, I do think there was a connection. A lot of things have been said about this and I don't think the real truth has ever come out. I don't know why. I suppose if the real truth came out a lot of people would be taken to task for not doing certain things they should have done. They were the higher echelons over whom neither the NCOs nor the men had any control.

I do believe there was some connection. I don't know why they didn't smell a rat. I wasn't with No. 2 platoon, I was with No. 1 platoon. I was sent out from FILTISAF, when they were killed, on the search party. I could never understand why they didn't take more precautions because the day before the ambush all the Balubas who were working in the camp – some of them worked in the laundry and others cleaned the camp dining hall, etc. – all demanded their money and packed in. Some of our people went to the village that night and there were no men in the village, only women and children. I think someone should have smelled a rat and taken more precautions. They probably would have saved some lives. The day before the massacre the Balubas had demanded their money and left.

The Baluba chief's son survived the shooting. He was in Albertville hospital for a number of weeks. We actually had a guard on the hospital while he was there. He was escorted back to his village by our people and handed over to his father. Lt Gleeson actually handed him back over to his father. He was okay; I think it was only a graze. The wound wasn't that serious. Naturally enough if it had been serious, being a head wound, it would have killed him but the bullet did not penetrate his head.

Personally, I don't think the Irish involvement in the Congo achieved anything whatsoever. The only thing it achieved was to make people in the UN understand that the Army here had people who could look after peacekeeping problems. That was because we never took reprisals against anyone. We were always ready to welcome and help the local residents anywhere we went. That was conveyed very strongly to the UN by the Congo mission. I think they realised that, as you can see from the missions they have done since then. We have had people in the Lebanon since 1978. They have done a wonderful job – not alone peacekeeping, but they've also done a PR job. They built schools and nurseries and have helped the people in every way possible. This is one of the strongest points on our side.

28. Joe Brady

JOE BRADY, DUBLIN

I served with A-Company in the 36th Battalion in the Congo. I was aged 17. We left Dublin on 5 December 1961 and returned in May 1962. When we arrived there the planes were being fired at. We stopped in Kamina overnight and we were then told to board the planes for Elisabethville. We were told to load up because we were going into a battle situation. The 35th Battalion was already there and under fire. When we were flying in, a couple of the planes were hit. We got straight off the planes onto buses and straight out into ditches and trenches. We were in the trenches for four or five days. We were coming under both mortar and rifle fire. A couple of our planes were shot at on the runway by Tshombe's forces, including mercenaries, who had guns on top of the airport roof at Elisabethville. We were shifted into a place called Rousseau Farm, it was a big horseshoe shaped thing. We were pinned down there. We went down a couple of times to try to take the troops from the tunnel. They were dug in along the tunnel which was the main entrance into Elisabethville. It was just a railway bridge and the traffic went underneath it. On 16 December we went out in force and took the tunnel. It was lashing rain. We were pinned down for a while. One lad – Charlie Raleigh from Wexford – stood out in the middle of the road and fired the recoilless anti-tank gun and knocked out the position. That is how we got into the town. They had the railway bridge and we couldn't get through because they were covering us all the time. We were pinned down in the hospital grounds. He took out the whole lot of Katangese troops there. We buried five of them afterwards. We advanced then right up onto the railway line and that is how we took over the whole town. The anti-tank shells actually went through the bridge. There was no fighting after that. Everything stopped then, when we moved into the town and took over. That was it, the whole lot finished.

We were not fighting alongside other UN troops. The Irish were there on their own. No. 2 platoon went up the main road and the other platoon went up the railway tracks. Lt O'Riordan and Andy Wycombe were killed that morning going along the railway. Our officer, Seán Ardan, brought us up through the main road. I think that Lt O'Riordan was killed by machinegun fire. When he fell, his radio operator, Andy Wickham, went out to pick him up and he was killed. That's how Andy died. A couple of days beforehand Corporal Fallon was killed in Rousseau Farm when a mortar dropped on the roof. A couple of other lads were injured.

I was not really scared or frightened. We had been well trained for what we were going out to do and we knew what to expect. I'd say some of the lads were frightened but you went out to do these jobs. I think the Irish UN involvement in the Congo achieved something at that time but I mean you are back in the same situation at the moment, aren't you? It's like the North – you get a bit of peace for a while and then it flares up again. It didn't really achieve a whole lot, in my opinion. You got a bit of peace for a few years and then it all flared up again. It is like everywhere else, one person wants to take over something: you get him out of the way and then someone else comes along. Look how many years we have been in the Lebanon. If the UN had stayed in the Congo for a few more years it might have made the country more stable. If they had policed it for a while it might have worked out a hell of a lot better.

DICK DUNNE, DUBLIN

I served in the Congo with the 34th Battalion from January 1961. It was a terribly frosty morning at Dublin airport. We left in the early hours. Boarding the aircraft was frightening, as I had never flown before. We were like bees strapped onto the side of this plane and down beneath us were armoured cars and vehicles of all types with tons of ammunition and weapons. We were just like flies stuck to the side of it. I was 32 years of age and held the rank of platoon sergeant. My platoon commander at the time was 2nd Lt Frank Coakley who is now a Brigadier General, O/C the Curragh. We arrived first in Tripoli and we thought the American base there was wonderful. They had everything – things we had never seen before. We were way out dress-wise because we were dressed in bull's wool, nail boots, and jam jar leggings. The first dose of heat we got was in Tripoli. It was sweltering, dreadful. We went around and saw all the fancy messes and dining places. The Yanks had things that we would not even see in posh hotels here at the time.

We then headed off for the Congo and were supposed to land in a place called Kano, but for some reason we did not land there. We spent an enormous amount of time in the air before eventually landing in Léopoldville at about 5.30 or 6 o'clock in the morning. We were marched to the back of some place. There was a coloured contingent – Moroccans, I think – which was responsible for looking after us. At that time of the morning we were presented with a big plate of rice with a big piece of steak sitting on top of it and a big bottle of hot coke. This was after spending about 16 hours in the air. The heat was humid even though there was no sunshine at that early hour. After some hours there we got back on the plane again for another six or seven hours and we landed at Kamina air base. We spread out there and got organised. We got into shape and had our daily duties set up. We familiarised ourselves with the set-up. We went out on various patrols to get to know the area.

We used to go to Kilubi which was about ninety miles out from the Kamina base. It was a powerhouse. We had a platoon out there on a weekly basis. The platoon was split in two: one half stayed in a villa which was up on the top part of it, while the other stayed down in a shed beside a river, which was right beside the power house. This was where Joe Flanagan, God be good to him, lost his two feet. They were blown off in a hand grenade accident. There was no one to replace him afterwards.

In Dublin I had received a vaccination that did not take until later, so I was quite unsteady on my feet. I was put in the front of a truck and brought off down to Kamina, and I took over from Joe. In fact, I was sleeping in the same bloody bed where he had been. Looking up at the galvanized roof I could see the hole that the grenade had made in the roof. It wasn't an enemy grenade. A lot of investigating went on involving the UN police but nobody ever found out exactly what happened. It was a home grenade, strangely enough, but no one knows how it got there or what happened. It would be remiss of me to say what we thought happened at the time. It would not be fair now, but it was an Irish Army grenade. We do not know how it happened, but the grenade got from the guardroom to underneath his bed. Grenades don't walk. That is the question, but we don't know how it happened. It is long since forgotten about but at the time there were a lot of to's and fro's as to what did or didn't happen. At the time the general feeling was that it happened as the result of a personal grudge.

We did duty at the airport which was about seven or eight miles from Kamina. There was base one and base two where we did airport guard. It was a massive area. Coming up to Easter we were told the Gurkhas were coming in. They arrived in force and we moved out. It all happened rather quickly. We loaded up and got out on Good Friday. There were all kinds of threats as to what would happen to us when we landed in Elisabethville because we were going to take over the airfield there. We were expecting all types of things to happen to us when we arrived there. Our planes landed one after the other on the runway and as soon as the planes stopped we came flying out of them like bees out of a bottle. The people who were supposed to attack and kill us were seen skirmishing through the bushes as we came out. We had no trouble and settled into villas there which were like bungalows. The Belgians had lived in the villas which were very fancy houses. They had left and had moved back to Belgium. The houses were left completely empty so we started taking them over one after the other. In some cases a platoon would share two bungalows, or maybe the best part of a platoon would be pushed into one bungalow. We camped in tents beside the airport and slept there.

We never came under fire there. There was some evidence of atrocities but I personally did not see a lot. Other companies had seen things. We had worked as train guards and saw various things but no atrocities. We remained in Elisabethville and were replaced by the 35th Battalion who had a hell of a lot of trouble a very short time after we left. We had a rather strict commanding officer, Eugene O'Neill, better known as the 'queen bee'. He had a strange set up. We had this war cry, 'Fala, Fala'. Each company was allowed a song they would sing going on and off parade in the morning when we were in Kamina. Ours was 'Twenty Men from Dublin Town'. At the time I recall that he [O'Neill] was from the Curragh and A-Company from the east were never too popular. We were supposed to be the gurriers, although that was far from being true. We pulled our weight with the best of them.

There would have been personal rivalries among the Irish troops up to a point. In certain cases, say, between the south and the east, there could have been little things. It would not have reached the point of settling old scores but it could arise when you had a get-together and a drinking session that remarks could be passed and fists could fly. There was a fair amount of drinking in the Congo. We had an old house in a factory and we used to get large bottles of Simba, a local beer. We used to go down occasionally to Elisabethville in the afternoon when things were nice and quiet. There were particular places you could drink in but other places were out of bounds and you were not allowed to go there. One particular spot was a fairly posh place. The Belgians – there were quite a few of them still there at that time – would buy a big bottle of Simba beer between four of them. The Irish guys would go in and buy a bottle each! About twenty minutes later they would order another four bottles and it would finish up with about twelve bottles on the table, as if it was going to run out of stock. Beyond that there was the occasional fisticuffs and what have you.

We got on very well with the Belgians we met. They had their problems, of course. The majority of them were anxious to get out because with Tshombe there at the time things were not the best for them. They were trying to get money. If we had dollars or sterling they were very anxious to get their hands on it and would give us a lot of Congolese francs for it because they wanted to move out. They would pretend to be going on a weekend trip abroad. They would leave the car at the airport and just bring a weekend bag, leaving everything in the house and take a plane. That was it, they were gone. That is how we moved into these houses some of which were completely furnished. The Belgians left in a hurry. Although they had been working there things were getting too hot for them. A lot of them were murdered but the gendarmerie and the mercenaries were there. Mike Hoare was one of the people there at the time who was in charge of a mercenary group. We had a couple of close runs with mercenaries and had some friction with them. I remember they threatened us at the airport. Ned Vaughan was our company commander and he had words with a mercenary guy. They tried to talk us down but it amounted to nothing – just threats and words. There were a few occasions like that where you would come across these people and they wanted to threaten you or tell you what to do, but we didn't take it.

I did not meet Mike Hoare personally. The mercenaries made a lot of money out of the Congo. We met a couple of guys out there – they were not mercenaries, they worked building the roads between the Congo and Rhodesia. One fellow was from Northern Ireland and another was Scottish. They made a lot of money. They worked on a system whereby if they were working in the Congo they got so much money paid in Congolese francs and so much paid in sterling. The sterling half would have to be lodged in a Rhodesian bank. That meant that if they had to run they had money to run to. Things were that desperate; people might have to run in a matter of minutes.

I think the Irish involvement in the UN mission to the Congo certainly achieved something. Had we not been there, there would have been wholesale slaughter. Even though we were ill trained and ill equipped, we adapted to the job. In our case, we took over from the 32nd and 33rd Battalions. We took on the job with very little experience and as time went on we picked it up and I think we did a good job.

29. Paddy Kirby

PADDY KIRBY, DUBLIN

I served with the 33rd Battalion in 1960. I was not supposed to go out. I was with the 3rd Battalion. But three men failed their medical tests with the Dublin battalions so they sent down to the Curragh for three replacements and I was picked to go. At first I was very excited about going. I told my mother and we had a bit of a party in the house. It was great. On the day of the party though, I was very sad about leaving. It was my first time away from home and I was very prone to homesickness. I had never been away before. Once we took off, all the fears went. We arrived in the Congo three days later. We went to Kamina straight away to guard the airport. Our main job for a while was to block the airport. A lot of Belgian army vehicles had been parked along the runway. Whenever a UN plane was coming in we had to clear the runway. We would drive out there and drive the trucks off the runway, and then block it again. The Belgians were in the airport at the time. During our free time, which wasn't much, we used to walk around and make a few friends. We used to drink with them in their mess. The Belgians said they were very glad to get out at the time. I got the impression they were very happy. They were a nice bunch. Not many of them could speak English but once you get a few drinks it is easy to communicate with someone. It's sort of universal. One thing I remember is that they were very young; they were even younger than us. We were only 18 or 19, but they appeared to be younger. Although I did not know it at the time, they were conscripts. We did not realise the situation we were in because very little information came down from the top. Day to day it was just a routine for us. We knew nothing of the whole picture.

I was on guard the night of the Niemba ambush. I was relieved that morning and went back into the billet and tried to get some sleep. It was very warm there, the windows had to be kept open. It was an ex-Belgian army barracks in Albertville. We had been transported by helicopter from Kamina. It was the first time I was ever in a helicopter. It was very frightening. I was relieved and went to have a sleep. That evening there was a hell of a commotion going on outside. We heard from one of the lads that there had been an attack. Sentries were posted all around. The guard was increased around the perimeter of our camp. We were there to defend a hydroelectric power station. We were terrified by what had happened although we didn't really see much. I did not see the scene of the massacre. A few Irish patrols went out all right. Comdt Crowley led one of them.

After the lads had been found they were put in wooden coffins. At the end of each coffin there was something like the handle of a brush sticking out. Apparently that was for carrying them. They were kept in a concrete building at the edge of the airport. I have a photograph of our platoon beside the Katangan flag. Behind us you can see the building where they were all laid out. What stood in my mind was that we had to mount a guard of honour around them. There was a hell of a smell. It was unbelievable. Then they brought these big oak, lead-lined, coffins over to prevent the spread of disease. It was very harrowing at the time and even thinking of it now it is very fresh in my mind.

As to whether the massacre could have been avoided, very little information came down to us from the top. Every day was a routine. We did not realise what was happening. We would go out on patrol but we did not know the geography of the area, although the NCOs and guard commander probably did. We did not know what the situation was with the locals. I do not know whether it could have been avoided at the time. To me, it was just a tragic accident. What really came to my mind was that we arrived there in the bull's wool uniforms which were very warm. We were in Kamina when the Belgians were pulling out, but they left an awful lot of ordnance behind; a lot of trucks, half-tracks and weapons. They left a lot of tropical uniforms. Our company at the time was supplied with Belgian army uniforms. All through the Congo I wore a Belgian army uniform. The only way you could actually tell the difference was that we had a UNO flash on our shoulder, but that was the only difference. I imagine that, to an ordinary local over there, we were just another regiment or company in the Belgian army. It is quite possible. I actually think that Niemba could have been a case of mistaken identity. Even the [Irish] officers over there were wearing Belgian army uniforms. We had no Irish army tropical uniform; it was Belgian army issue. The 11-man patrol at Niemba was wearing the same as I was. We had shorts and long combat trousers. We were issued with boots, which was brilliant because we had never seen rubber boots before. We always used to have the hobnail boots. Even to this day I always say that maybe Niemba was a case of mistaken identity. It is just like when anyone goes out on patrol, it is a risk you take. You never know who is going to attack you for any reason.

Myself, Mick Monaghan and Michael Colton and six others went off in two jeeps one day. We drove out about three or four miles into this village. We were looking for souvenirs, but apparently some of the locals were getting very agitated at the sight of us being there. They actually thought we were Belgians. We left our weapons in the Land Rovers and luckily enough there were two lads in the Land Rovers. They spotted us and the first thing Mick Monaghan said was: 'Lads, get out of here quick.' The jeeps had been turned around. Two days later the lads were attacked at Niemba. We did not know. We just went out there at that time, without knowing the local agitation against the white man. To them every white man was a Belgian. An old coloured man was walking down the village with this antiquated blunderbuss. It was a huge yoke, an elephant gun or something. He was being held back by the rest of the tribes people. This village was within six miles of the power station. We thought it was just a holiday camp. We drove around to the village to make ourselves known but apparently any white man was classed as a Belgian. This was two days before the Niemba massacre which, quite possibly, could have been linked to the wounding of the Baluba chief's son. At the time, we had not heard about the Baluba chief's son having been wounded, but from where we were it looked hairy. We scarpered back and later on we started worrying about it. We had left our rifles unattended in the village, but you learn. A few people were agitated at the fact that we were there. They did not want us there. We were so naïve we thought everyone would love us.

In addition to the Belgian uniforms, we had plastic helmets, like American ones. They were blue with UNO on them. That's all we had. The Belgian troops wore berets, coloured red or black. In the heat we used to take the plastic helmets off. They did not afford protection against anything as they were made of plastic. Once you had the plastic helmet off – most of the trucks were lined with these hats – you'd look like a Belgian. A lot of the trucks were left over from the Belgians. Most of them were painted white with the letters UNO on them. The trucks were very identifiable but that didn't register in their minds. They thought we were just another branch of the Belgian army. The UNO letters were the only distinguishing features. The troops that went out later on were well equipped to eliminate that.

With hindsight I don't think anything was achieved by the Irish army going to the Congo. It is the same situation there today. [President] Kabila is there at the moment and there is anarchy there again. When Mobutu got into power he was more or less a despot himself. He was no better than Tshombe. The only real saviour of the Congo was Lumumba, but he was branded a communist and was executed by Tshombe. Given its wealth, the Congo should be one of the most prosperous countries in Central Africa, but the money and natural wealth was squandered over the years. They never let it filter down to the people. Even today, there is a war there and at the moment Kabila is getting hammered by the Rwandans. It is really tribal over there and always was.

30. Michael Phelan

MICHAEL PHELAN, DUBLIN

I am originally from the Curragh, Co Kildare. I passed out from the Depot after six months training in May 1960. In November 1961 they were looking for volunteers for the Congo. I had just turned 17 when I volunteered and was looking forward to this adventure. I arrived in the Congo on 19 December 1961 with the

36th Battalion. I was stationed with the anti-tank section in our headquarters company. I was one of the youngest members of the Army to go to the Congo. My first recollection was of going on the Globemaster aeroplane. I was one of the people selected to guard ammunition. There were only twelve or thirteen of us on the plane and throughout the journey we were sitting on ammunition boxes. When I arrived in the Congo the first thing that struck me was the heat. It was very, very warm, but as I was small and slim it didn't affect me that much. We arrived at a transit camp in Léopoldville and most of the other UN troops there were Pakistanis who were all drivers. I was there for about eleven weeks and the food was absolutely horrible. I had a ferocious appetite but strangely I was one of the few people who did not get the runs. I do not know why. In the first weeks the lads were all taking turns to sit on the latrines. But I was one of the few who wasn't affected and I was able to eat what I liked and nothing happened, thank God. The lads were slagging me about it.

The food consisted basically of pancakes, powdered eggs and powdered potatoes. I didn't see any meat at that stage. It was horrible stuff but, before we went out there, the Army grub wasn't great either. It didn't bother me. I was always hungry and I ate it. In Kildare Barracks you could hit the black pudding against the wall and you could catch it when it came back to you. We used to get a loaf of bread and it would be cut four ways, with a little bit of butter on top.

In the Congo, my first experience with the natives was when I looked up into a tree and saw a young lad. He must have been crouched there for ages because he had caught a bird and had it by both wings. I presumed he was going to eat it. When we had finished eating our pancakes and powdered potatoes we came out and there were bins to scrape the leftovers into. The kids were actually in the bins. You could see the sores on them and the flies and wasps on them. They were trying to eat what we had thrown out. So, after that, I never complained about food. It was a great education for me to see that at that age. The memory has stayed with me to this day. Due to that, I have always appreciated how well off we are here. I was very young and did not know anything about the political situation in the Congo, but I remember that one in four of the local children had malnutrition, swollen stomachs.

The battle of the tunnel in Elisabethville had taken place only a few weeks before I arrived in the Congo. I did not see any action there. There was a lean-to where Mass was celebrated and at night they used to put beds in it so that people could sleep there. A young fellow called Fallon was sleeping there when the building was hit by a mortar shell. I think it was fired by the Swedish troops, but I am not sure about that. The mortar dropped short and a wooden beam struck the young chap on the bed and he died. Thank God, I was not in any accidents, but I had a few scary moments. For instance, I was coming into the camp one day in the farm in Elisabethville and I kicked a rock over and there was a snake under it. I do not know whether it was deadly or not but I ran like anything to get away. The lads shot at it with Gustav machine guns and blew it to pieces.

On another occasion, after the monsoon rains, holes would appear in the roads so we would have to get material to fill them. Myself, the sergeant and four or five lads went to a brick works near Elisabethville in a jeep which we filled up with bricks. On the way out the owner of the brick works followed us in his MG car. The sergeant was driving our jeep and sped around the corner so I was left on the ground. The man in the MG put me in his car and drove me to the camp. I was glad because I wouldn't have been able to find it. When we got into the camp he said: 'You are going to show me the adjutant's office', but I just legged it away.

I met Conor Cruise O'Brien in a transit camp at Léopoldville. I was going out the gate one day and I recognised him as a man in the media. I said hello to him. He looked great, sauntering around in his dark glasses and he had nice gear on him, including shorts. A while after that word came through that Dag Hammarskjöld had been killed.

I recall that after flying from Léopoldville to Elisabethville – a journey of 1,000 miles – we had to cart the ammunition off the plane. It was hard work but the lads were great comrades. We thought immediately that we would be going somewhere to eat because we were starving, but the sergeant said: 'No lads, you're on patrol.' It was called a chain patrol. Katanga was very big, so you would have to walk a mile and meet someone else who would walk on further. I realised when it became dark that I wasn't meeting people that often. The chain was being broken. That was the first duty I was on. It was a bit scary and I remember saying: 'Jesus, will the light ever come?' I remember after that first night, at five o'clock in the morning, I was delighted to see the daylight coming and to get to the end of that duty.

I will always remember the camp in Elisabethville where there were over 3,000 Balubas. When the civil war started some of them were cut off in Elisabethville while others were in this camp near the farm. One of our morning duties was a water patrol. We had to organise the queues for the women to get water at four o'clock in the morning. It was all the poor women, with babies hanging out of them everywhere, who got up. They would have basins on top of their heads for the water. I don't know where the men were. They must have been sleeping it out. It always struck me how hard the poor women had to work. Our job was to line them up to get to the water. There would be mayhem because they all wanted to get so much water to do their washing during the day.

One night in that same camp something happened and I saw the military police patrol chasing after a black chap who was running through the camp. It was pitch dark. The conditions we in the Army lived in were poor but when I saw those poor people I never complained again about the food. For instance, they lived in one-room huts. When they were looking for this chap, I went around with them and saw whole families like a pack of cards piled on top of each other in that kind of atmosphere – the extremely hot weather. Those people lived in very bad conditions. The Irish were good to them. We used to get our meals delivered to us by jeep, but we would immediately share the meal with the young coloured kids as well as the fathers and mothers – whoever wanted some food. The same thing happened in the dining hall in Elisabethville when we went for our dinner of pancakes. I didn't see any meat there, although some of the lads told me they got meat but I don't know where they got it. We would get twelve or thirteen pancakes with dog biscuits. You would have to use the butt of the rifle to break them. You would see the kids' noses flattened up against the window so we would go out and share whatever we had.

I was in a villa in Elisabethville with the lads from Kildare Barracks at first. I was a Bren gunner and was only about five foot, five inches tall. They used to call me Matoto which was Swahili for baby soldier. I had to go to B-Company and take over a Bren gun so I was well armed. Carrying these two machine guns around, the heat was amazing. I had a great time with the lads of the Southern Command. There was a cook there who cracked up one night. When we got this villa we had to try and pick positions for our bunks, but I was unlucky and had to sleep under a hole in the ceiling. Every time it rained I had to pull my bed away from the hole. One particular evening it was raining and I woke up. There was this cook who was in trouble and went missing. I waited a while to see if he would come back. He said he was going up to the Baluba camp. I went in and told the C/S whose name was Timmy Fennell from Limerick. He asked me when the cook had gone and I said it was about fifteen minutes ago. As it transpired, he went up to the Swedish camp and was actually shot at. He was brought back and the military police came and took him away. I heard later that he had been sent back to Ireland.

We had young black kids and men working with us. They would wash and iron the clothes for us, but I heard that the Swedish soldiers would not let them into their camp. It was different. The locals used to cook for us but the Swedes would not even let them into the camp. Therefore, when the Irish cook went down there at night shouting and roaring they just shot at him. That is what happened to him.

Overall, it was an absolutely brilliant time and I made great friends there. Looking back on it, I think the presence of the Irish Army in the Congo most certainly served a purpose and did some good. First of all, you have to remember that when we went there we were green, to say the least. We even brought old army uniforms with us. We did not have the greatest gear, although my battalion had the white, light uniforms. The green uniform came in handy because it got very cold out there at times. The Irish shone in their communication skills. They had this gift to get on with people. Even at a young age I noticed that they didn't have any pre-conceived notions about anybody. If there was any trouble anywhere the Irish would be the first in communications skills to go in and sort things out. It was the same in every other post I went to throughout the UN. That was the strength of the Irish. The Irish troops were very highly thought of and did a great job. It was a hard job.

The local people always got on well with the Irish troops. There were no rules against them coming into our camps or villas. The Irish lads bought souvenirs from them and they shared their meals with them so there was no problem whatsoever, that I could see. The conditions were hard for the lads. I was only a young single fellow but it was hard on the married men with families at home, who were on active service. When we went to the loo, the gun was loaded because you never knew if something might happen. I was lucky enough and nothing happened to me, although terrible things happened to other people. I saw terrible things and you are under stress all the time. A married man with two or three kids would always be under more pressure.

The most horrible things I saw were the conditions those people lived in; the lack of food and medical care. I was too young to realise about the lack of education but the school system was not great. They just had nothing, they were living on their wits. I would like to see what it is like now, but the conditions were very hard then. I did not see any massacres but I heard about Moroccan or Indian troops who died when an anti-tank gun was fired inside a house. One guy was messing with a loaded anti-tank gun, he fired a shell and brought the house down. Seven or eight fellows were killed.

After the civil war there were not many Belgians left there. We used to go into Elisabethville and you would only see a few shopkeepers. You would never see people on the streets, as in normal times. There was a fight for control of Elisabethville and consequently the post office and other buildings were riddled with bullet holes. It was a good day out to visit Elisabethville where we could relax. On the way we would pick up with our units along the roads. We would walk around, have a few drinks and take some photographs. Most of the Belgians had left by the time we got there so we took over their villas. But you could just see how the black population lived under the regime. Outside the villa we lived in was a hut like a coal hut. There were chains in it and seemingly this is where the natives slept who worked for the Belgians. There were old tin plates lying around where they had been fed like animals. They had no love for the Belgians, Italians or any other white people who had been there. I was told that the blacks had been chained up in these horrible conditions.

During twenty-two years in the Army I was never marked absent, but I was in the Congo. It was because of my size; I was very slim and only eight stone in weight then. One day the officers were looking for a patrol and they came into my room. They looked in and thought I wasn't there but I was asleep under the blankets. They didn't see me so I was marked absent. Later on I was asked why I had not been there to report for duty, and the lads had a great laugh about it. I am famous in the Army for that incident. I thought I was one of the youngest soldiers in the Congo, at 17, but I heard that there were two 15 year olds there too. If you could get a note from your parents to join the Army and you were big enough they would take you. I was a gunner, trained to fire anti-tank weapons. When we were practising on the ranges there was a big fellow from Wexford who was supposed to be the number one gunner and I was the loader, but he couldn't hit a barn door at fifty yards. I was able to hit the target with the first two shots, so I was the number one gunner on the 84 recoilless rifle. I did not have to use it in the Congo; it was in the store all the time but, if needs be, I could have used it. I also had a Bren gun and an FN rifle, so I must have been like Rambo.

TONY NORRIS, DUBLIN

I think every man in the Army volunteered to go to the Congo. I volunteered for the fist two trips – the 32nd and 33rd Battalions – with no luck, but I went out with the 34th Battalion eventually as a Private, with my brother Frank. He is dead now, the Lord have mercy on him. That was in 1961. We served out there together. He was a cook. While in the Congo, I was promoted to acting Corporal. The same thing happened when I was with the 37th Battalion. The two battalions were different units altogether. The 34th Battalion was all heel ball. Colonel O'Neill was our commanding officer; we used to call him the Queenie O'Neill. We all had No. 1 haircuts and had to wear our vests during inspections. He was a good officer though, and he kept the morale going but he was a stickler. You would have to present arms about fifty yards before he came by you, and if you looked left or right after he had gone by, you would be relieved of your post. That was how strict he was.

We were based in the airport at Elisabethville. There were also Gurkhas and Swedish troops. The Swedes gave us a bit of trouble at first until Col O'Neill got on to them. They used to go around with these specially trained dogs, which would bloody well eat you, oh Jesus! We would be on different posts and they would come on patrol, walking by you with the dogs. The dogs would go for you, but they had them on leads. They were bloody big, huge mixed breeds. I have never seen dogs like them. All the dogs were shot before the Swedes left. The dog handlers could not take them back home. O'Neill warned the Swedes: 'If you don't keep those dogs away from my men, I'll shoot the dogs myself.' That kept them off our backs.

On another occasion things got a big nervy. The gendarmes were coming in to take the airport from us. They said we were surrounded for three days and nights. We could not get a wink of sleep apart from catnaps. It got to the stage where you'd ask them to come in and take the airport. We were browned off waiting for them. The mercenaries and black gendarmes wanted to take over the airport but O'Neill warned them that if they came in he would wipe the whole lot out. We had a platoon of Gurkhas out in front and all our lads were in trenches. I would have been the last to be got because I was up in the control tower with a Vickers machine gun – myself and another chap called Moore. Thank God they didn't come in. They got the message and they moved off. That was the nearest thing we had to trouble.

When I was serving with the 37th Battalion we found a few bodies here and there. They had been shot. I picked up a helmet at one stage and there was half a head in it. It sickened me, it really did. We were on guard holding three mercenaries prisoner in a small room. We had a sentry walking up and down outside and another man sat inside with his back to the door with his rifle pointing across his lap. Another of our men was in the corner. We used to let the prisoners out of the cells for a bit of air. They used to pretend they couldn't speak English, but we would not give them an inch. We did that guard duty for a week; it was nerve-wracking watching these fellows. One of the mercenaries was Belgian and another was a Frenchman. The Nigerians relieved us, and about an hour after we left we were all called out. When the Nigerians took over guarding the mercenaries, an officer called Jack Phelan heard the shooting. He went back up and found three Nigerians dead. One was nearly cut in two. The three mercenaries had gone. Seemingly what happened was that one of the Nigerians took it very cushy and put a submachine gun on the table. The mercenaries were up like lightning and disappeared into the jungle. They were never found again.

We did a good job out there. The Irish troops were respected by everyone. They brought a lot of peace then, but it is now as bad as ever out there. They did a great peacekeeping job, definitely. A lot of people came back suffering from post-traumatic stress. The way I see it, myself and my brother and our gang were made going out anyway. It didn't suit everybody. A few mates of mine broke down when they came back. One of them used to be in Grangegorman. He would not even recognise me when I'd say hello to him. He is completely gone, which is sad. The heat got to a lot of people, they weren't able for it, and the food was crap. You were eating flies and everything. With the 37th Battalion we ate bully beef every day, for breakfast, dinner and tea. The grub was bad. We had a good trip and we survived. We were issued with blankets which you didn't need out there so we gave them to the villagers in exchange for local beers called Simba and Primus. (Simba was the local brew; it was deadly. You'd put the bottles on the ground and they used to blow up like hand grenades in the heat. One fellow got cut under his eye by a piece of glass.) Then the military police came down and saw that there wasn't a blanket left in A-Company of the 37th Battalion, so they invaded the village and got all the blankets back, but they were all gone again that evening, back to the village. It got to the stage that the military police would not come into the camp, especially at night-time. We were out drinking and enjoying ourselves.

We would be put out in a mud hut in the middle of the jungle, doing guard on post. We had tilly-lamps. As soon as you lit the lamp there were dragonflies and mosquitoes flying around. You could not stay in the mud hut or you'd be eaten. I used to leave the hut and stay under a tree in the dark. It was ridiculous being sent out there on your own, about 600 yards from your guard post. I was often left for four or five hours on post because it might have been too windy for the next guard to relieve you. It was great to see the dawn coming up. I got bitten by mosquitoes but funnily enough I was not affected by them as much as

my brother was. He was like a leopard because they used to take to his skin better than mine. They gave us this [insect repellent] lotion and the smell of it was
unbelievable. When we put it on the mosquitoes thrived on it! Before going out
on patrol we would cover our beds with mosquito nets, but when we had gone
the lads would lift the nets and let all the mosquitoes in. When you got into bed
you'd be eaten, so the lads would be laughing and having great crack.

On my first trip I didn't eat a dinner for about three months because when
you got dinner all different types of flies would come in like dive bombers. We
used to cut a large square of mosquito net and when we got our dinner we would
eat it under the net. That is the only way you could enjoy it. You'd take up a
slice of bread and there'd be little black flies baked into it. By the time you had
finished taking them out the bread was like a sieve. It got to the stage where you
would just butter over them and eat the lot. They were baked so it did not make
any difference.

I met a few Belgians out there. A young mercenary used to drink beers with
us. He bought us a few beers, which was great. We would order a bottle of
Simba beer each, but the Belgians would share one bottle between five of them.
The blacks couldn't believe it and they were delighted to see us coming in. We
would not share a bottle. The young mercenary was about 22 years of age. One
day we met a mate of his who told us he had gone into the jungle and never
came back; the blacks got him in the jungle and killed him. That's what they
were up against all the time. The mercenaries were working for the Katangans.
You didn't know who was who among the blacks. The Moroccan and Tunisian
UN troops were out and out dirty bastards. We went to take over a billet and
there was a line of lockers down the centre of the room, which they had used as
toilets. We had an unbelievable job cleaning out the place. I didn't like the
Moroccans and Tunisians at all. They were a clan on their own and didn't mix
with anyone. We got on well with the Swedes but they lived high and dry. Their
mortars would be mounted and they would be sitting down eating cakes. They
had plenty of money and were better paid than we were.

The Congo was all muck, dead heat and humidity. When you came back from
patrol you'd think you had a great tan, but it was only the red dust from roads,
which washed off in the shower. I served in Jadotville and Elisabethville. The
37th Battalion was much rougher than the 34th – they were all head cases in our
company. That is what kept the morale going. My brother, Sammy Greegan and
Ali Duff were all drinking buddies. They are all dead now, Lord have mercy on
them. I do not think Katanga could have been taken under UN control without
force, not with all the mercenaries in there. Force had to be used. Unfortunately,
we lost men out there which we should not have done. When the lads of the 36th
Battalion took the tunnel in Elisabethville, the Swedes were supposed to go in
with them at dawn but they didn't turn up, so the lads went in themselves and
took it. The Indians had it and lost it. They just couldn't hold it. There were so
many nationalities out there, including Pakistanis and Indians.

ALPHONSUS FRANKLIN, LIMERICK

I went out to the Congo with the 33rd Battalion in 1960. We were based in Manono, a big tin mining town. We were responsible for protecting the miners. We were there for about four and a half months and then we went to Jadotville for the last six weeks before coming home in January 1961 with the 32nd Battalion. Most of our work concerned patrols and protecting the miners. I remember the day when we heard about the Niemba ambush. I happened to be passing a patrol that went out to Niemba from Manono. Needless to say we met all the other patrols from the rest of the companies. I was there with the people who found Private Kenny and Private Fitzpatrick the day after the ambush. There was a third survivor called Private Browne but others can tell you more about him than I can. I remember that when I went out to the Congo again with the 38th Battalion in November 1962, we were told that Private Browne had been found by missionary nuns there. He had died and his body went back in one of the planes that brought the 38th Battalion out. Afterwards I found out that this was true.

The commanding officer of the 38th Battalion was Col Delaney who later became Chief of Staff. But he only remained in the top job for about 90 or 100 days before he died. He became director of intelligence at the time of Charlie Haughey and the gun running, and all the things that happened in the North of Ireland when he [Haughey] was put out of the government.

We saw limited action. We left Elisabethville for a place called Kipushi on the Rhodesian border. During our journey we came under fire on ten or eleven occasions from the Katangan army which was led by Belgians. There were even Swedish mercenaries there. When we came under fire adjacent to the zoo in Elisabethville, where we had to take cover, one of my friends shot a sniper out of a tree. He happened to be Swedish and very young, about 18 or 19 years of age. Personally, I did not shoot anyone – at least, I don't think I did. We had big problems getting to Kipushi, although later – within three or four weeks – it became a haven, a real holiday camp for us. We went to Jadotville where we stayed for about three months. I had a bad accident there and was in plaster for that time.

What I saw the day after the Niemba ambush will never go out of my mind. In actual fact, we brought the two vehicles the patrol had at Niemba back to Manono. Baluba signs were painted all over the canvas of the vehicles with the lads' blood. This was proved afterwards. The lads were in bits. I do not want to go into it any more than that. I will let somebody else tell the gory details. I saw a bit of it but I did not see too much of it. We brought the vehicles back to Manono and I have to say they were bad.

I think the Irish involvement in the Congo was marvellous; it made the Irish Army. We were second-class citizens before that. Nobody wanted to know anything about the Army but that attitude changed after the Congo, and especially Niemba, when people at home realised that we could be killed for what we were doing in the name of peace. If I were asked whether I would go again, the answer is yes. It was a fabulous country. The Belgians were affluent, they had everything that was going, whereas the natives were downtrodden. This is where the Irish succeeded more than anybody else. We were able to look after the less well off in the Congo – the black people – and they took to the Irish. What happened in Niemba was a one off thing, and it never happened again. I am sure every fellow that was out in the Congo would go back there again. To my mind it was the best thing that ever happened to the Irish Army. I have been in Cyprus, Sinai and the Lebanon, but the Congo was the forerunner to all that.

Some funny things happened in the Congo, like when we were having a singsong in the canteen and the lights went out; when the lights came back on my beer had gone! We enjoyed ourselves, even though we did not have any place to go at night apart from the canteen. There were no discos or cinemas to go to. From the time you left Ireland until the time you came back, your whole life was completely and utterly involved in the canteen or the officers' mess, whichever you belonged to. Your social life revolved around that. Nowadays, no matter where they go, the troops do everything that everybody else does but at that time there were no holiday weekends or home leave.

Two Limerick lads – friends of mine – were killed. My company sergeant, Felix Grant from Clonmel, died in Manono. Billy Bolger died and Jack Geoghegan, from County Limerick, was shot by one of our own. I was down there with his family recently laying wreaths on his grave.

When we went overseas we started to get to know the rest of the Army. Before that we did not know anyone outside our own barracks. Companies were drawn from different parts of the country. Battalions were from the West, East, South or North. When you went out to the Congo you got to know people and those friendships have lasted up to today. The Congo was the making of the Irish Army, without a shadow of a doubt. We have never gone back since. We were backwards at that time; we were barrack room soldiers. We would go on parade, polish our boots and lay out our kit. Now they don't have time to lay out their kits; they just get ready to go overseas. Yes, the Congo was the beginning of it all.

BARNEY RING, CO LIMERICK

I went out to the Congo in 1963 around the time of the first Irish racing classic in the Curragh. All my memories about going out there are of hardship from start to finish. Being transported in the back of obsolete transport vehicles for long distances and not having any facilities when you got there. I was a mechanical fitter and we didn't have a workshop so, being the 2nd Infantry group, we were literally the last such grouping there and we weren't replaced. Most of the vehicles were antique and beyond repair so we were left pretty much with a load of scrap and we had to work with that. I do not think I'm exaggerating when I say that we had to work from approximately seven o'clock in the morning until evening time, otherwise nothing moved. There was a siesta at lunchtime because it was too warm to work outside. As a fitter I pretty much had to be a jack-of-all-trades and you had to keep all the available equipment moving wherever possible.

I did not see any military action there but we took precautions. We were armed twenty-four hours a day and we ensured that the arms were not too far away from us at any time. When we were going out, the weapons were certainly always with us and fully loaded.

We were based in Kolwezi, which was not too far from the Rhodesian border. We did patrols along that border. We were there at a transition time when the Belgians were moving out of the place. They were being forced out by the natives so there were only a few remaining that I knew of. I can recall the last Belgian getting out of the place and we pretty much escorted him to the Rhodesian border and took him across. We had pretty good relations with the Belgians because we were the only people they could depend on. Remember that they were in isolation at that stage. They had been the invading force and had taken every bit of wealth out of the Congo – mining resources and whatever else was there. Malachite would have been the main source of mining, but there were also a limited number of silver mines. Malachite is a rather expensive item used in the making of jewellery.

The Belgians left with us because we were the last to leave. I do not think any foreigners remained unless they were in the diplomatic corps. Conor Cruise O'Brien would have been amongst them at the time. I saw him there but I did not have a chance to talk to him, not that I'd be particularly interested in doing so. He was based in Elisabethville about 200 km away and he had an Irish driver.

While we were there we served a useful purpose but, on reflection, I thought that when we left there we hadn't achieved much. It would possibly have been more realistic to stay there as a temporary force until everything was resolved. History dictates that everything wasn't resolved there. You would have to consider the purpose for which you had been there in the first instance, because you didn't achieve much by the fact that you didn't finish your business. It is difficult to organise things when you are going into a territory that you know little about. When we got there we had to organise ourselves. We had the slight advantage of having people who had served there previously, and who were able to help us by educating us about the terrain and the dangers involved.

BILL WHELAN, CARRICK-ON-SUIR, CO TIPPERARY

I went to the Congo in August 1960 with the 33rd Battalion, and I later returned there with the 35th Battalion. I remained there with the 35th for a couple of months. There was trouble in a port called Matadi. There were Sudanese troops there along with Canadian signalmen. They were disarmed and sent back to

Léopoldville so all the UN supplies were tied up in the port. General Seán McKeown looked for volunteers from the Irish contingent to go back in there and I was asked if I would go. Four of us went back in unarmed with our uniforms on. We were arrested because they thought when they saw us that the UN troops were going back in. After negotiations it was agreed that we could stay there to receive the port supplies provided we did not dress in uniform. About six months later I went back out to Matadi on my third tour of duty.

I did not know the men who were killed at Niemba. We were in B-company, most of whose members were drawn from Cork and Limerick. Those who died at Niemba were drawn from A-company, mostly from Dublin with a few from the Curragh. We were quite near them but we didn't know them.

We came under fire with the 33rd Battalion, although not anything as bad as it was during the heavy fighting in Elisabethville, in 1961. We did not see anything like that. In Matadi, we worked with the Belgians so it was friendly enough. But with the 33rd Battalion when we were in uniform we found they were a bit two-sided. When the chips were down, however, and they were in trouble, Comdt Patrick Pearse Barry evacuated approximately 240 of them from a place called Manono. They were in danger of being slaughtered so he got them out of the country. I don't think he got any thanks for it. They were out and out businessmen. They were airlifted out from the very small airport in Manono, which was the headquarters of B-company, the southern command of the 33rd Battalion. Tony Connolly, who was later Quartermaster, was the signalman there.

There was a good bit of fighting between the natives. They could not seem to agree. There was all sorts of hassle. We were at a small outpost called Piana Mwanga, a few miles from Manono. Once a week we travelled into Manono to get supplies. Comdt Harry Goldsborough would say: 'You have some passengers this morning.' They would be wounded natives who had been fighting among themselves and who had to be brought to hospital. They'd be put in the back of the Land Rover to Manono. They were suffering from bullet wounds and some arrow wounds, although the arrows were not still in them.

I certainly feel the involvement of Irish troops served a purpose in the Congo. We did very good work out there. It could be said that the first group that went out was not properly prepared, although they went out with good will and they did their best. They were not properly clothed or armed. We used to travel from Piana Mwanga to Manono by jeep. I was a corporal; we had a driver and another private. The three of us were in a very small jeep and it was only afterwards that we realised we had been sitting ducks. After Niemba things began to tighten up and we did not take any more chances. Looking back on it, I think we were very lucky to survive the Congo. We used to go to Mass every morning and every evening and that helped us. Niemba is an example of what could go wrong. We went out with the best will in the world and thought everybody would be friends. After Niemba, however, people began to say that we could not take any more chances. People blamed the troops at Niemba – but I certainly wouldn't blame them – for not being more aggressive. That wasn't our policy, but they paid for it. Looking back, we could have been in the same position. It was very dicey if you showed too much force. For instance, when we went to Mass I was carrying a Gustav sub-machinegun, but you would hide it, not display it. I suppose the same thing happened at Niemba where they did not want to show force. They wanted to show they were there as peacekeepers, but they paid for that.

It was a very difficult situation to go into a civil war like that. Even if two people are arguing and you step in and try to make the peace, you have to be very diplomatic. It is not an easy thing to do. It is a very gratifying job if it comes off all right. You could end up being dragged into the argument, however. The Irish were noted for being diplomatic. They'd shake hands with anybody and didn't care who they were. Even if a fellow was prepared to stab them in the back, they went over and shook hands with him, and that helped them an awful lot. Of course, it worked the other way too.

After Niemba, relations between the Irish and the Balubas were very hostile, naturally. We heard stories of Irish soldiers who, maybe after a pint or two, said they wanted to go and get these guys, although I am not saying it happened. That's a natural reaction. It is like the situation up in the North of Ireland; if someone kills your brother you feel like having an eye for an eye, but there was no retaliation after Niemba. I was on duty as guard commander at the power station in Piana Mwanga. At about 5 o'clock in the morning we got a phone call from Harry Goldsborough. He said: 'Bring up the members of the guard, I have something to say to you.' We went up to the bungalow where he was and he said: 'We have some bad news from Albertville. There has been an ambush at Niemba and some of our people have been killed.' It was a terrific shock. I remember that day. I had been up all night and the previous day, and I didn't sleep. A Baluba child came out and waved. They would give you little gifts. She handed me an egg and I felt like hitting her with it. I counted to nine, said, 'thanks very much' and shook her hand. But it was a hard thing to do. I could see the father in the background with a scowl on him. They knew what had happened because Niemba was only a short distance away. It happened and there was no point in reacting to it, although you would feel like reacting.

There was great hunger, even famine, there. One day I was walking down a street in a small town with a friend of mine who was a corporal. There was a woman sitting on the footpath with a couple of fish for sale on a mat. My friend nudged me with his elbow and said: 'Look at that.' I looked and had to half close my eyes before I could see that the woman had two dead rats also. She was selling the rats and the fish. In broken Swahili I asked her how much and she said five or ten francs. She asked me if I wanted one or two, and I said I'd buy two. I thought to myself that I would create a bit of history [buying the rats] because it had never happened to a Carrick man before and I could always boast about it. I paid her the money and I said I was going to town and would collect them on the way back. Like the Three Wise Men, however, we went back to our own country by another route and I didn't collect the rats. I am sure she was rubbing her hands saying: 'The stupid Irishman paid but never collected them.' At that time, in late 1960, the local people were reduced to eating rats.

In 1972, I went back out to Africa as a civilian volunteer, working as a part-time woodwork teacher in a school in Sierra Leone. I thought it would be different there as, although it had been a British colony, it was independent. One morning there was a commotion in the school. We had bags of meal they used to cook for the 11 o'clock tea. I asked what was wrong and they told me: 'A rat, a rat!' Like a brave Irishman I stood back and let them at it. Later on I noticed this boy was sulking and crying. I asked him what was wrong and he said he was upset because another boy had stolen his rat. Earlier on I had given him a shovel to bury the rat. I saw the rat lying on the windowsill and they told me that they would bring the rat home to make soup. I learned more from those young teenage children that day than they did from me. When I had seen the woman selling rats in the Congo I thought it was unusual, but ten years later I saw the same thing in Sierra Leone. The locals had very little protein, so anything like rats or monkeys were eaten.

In the aftermath of Niemba, there were some Baluba prisoners in a hospital in Manono. A medical corps doctor, Comdt Arthur Beckett from Cork, was working with them. I brought some wounded people in there but whether they were Balubas from Niemba or not, I don't know. When we left, Comdt Beckett got a radio message to stay in Manono to collect the wounded who were being taken to Albertville to be tried. The message was sent in Irish to Comdt Beckett so that the Belgians would not understand it. They translated the order to stay in Manono, as 'Fan sa fear ó ní headh'. Whoever thought up the translation deserves a medal because it would stump anybody. The incident was described in Tom McCaughren's book, *The Peacemakers of Niemba* [Dublin, 1966, pp. 44, 98].

SULLIVAN McSWEENEY, DOORADOYLE, CO LIMERICK

I went to the Congo in August 1960 with the 33rd Battalion, and I went back again with the 36th and 38th Battalions. We knew very little about the Congo before going there, apart from geography lessons at school. It seemed to be a very hostile environment to go into compared to what we were used to. The way people lived there was completely alien to the way we understood normal living to be. Generally, however, we found the local people to be friendly and helpful once we got over the initial language problems. We learned a bit of Swahili so we were able to converse with them. Our army training was not of great use to us initially, but the training we received on the ground out there made things much better. As we had not served abroad before, we found it difficult to set up proper dining facilities, cook houses and washrooms.

Three of the units I served with were involved in some of the troubles. Men of the 33rd Battalion were involved in the Niemba ambush in November 1960.

Nine out of eleven people on that patrol were killed. We were sent out on patrol looking for them when it was discovered that they were missing. We found some of the bodies and put them together in canvas sacks and sent them back to headquarters in Albertville. My company was stationed in Manono which was an outpost, while all the other companies were in the vicinity of Albertville – just two or three miles from one another. We were about 200 miles from the rest of the battalion. We had a few skirmishes but a lot of it was internal fighting between the Baluba and Songe tribes.

I do not think the Niemba ambush could have been avoided because, from things we were told, the Belgians were in the process of pulling out of the Congo at the time. The Belgian paratroopers used to wear a type of blue helmet. Apparently the Congolese had suffered fierce pressure under those people and naturally they resented them very much. Given the fact that the Belgians were pulling out, the local Congolese saw an opportunity, thinking the Niemba patrol was Belgian. It was the first patrol into that area where these people were literally living in the bush. They were armed with old blunderbusses and bows and arrows with poisoned tips. I think they understood that the Irish troops at Niemba were Belgian paratroopers and when they saw the opportunity of having a go at them, they did. The Irish fellows were in the wrong place at the wrong time. Even a month later it most certainly would not have happened. I say that because word of the work the Irish were doing had spread rapidly around the Congo and they were held in such high esteem that this would not have happened. But then again, if it was their first encounter into this area, God only knows, they might still have thought they were Belgians.

The Balubas used to dip their arrowheads into poisoned herbs. If you were struck by an arrow, that poison would travel rapidly through your blood stream and kill you. The arrow wound itself did not kill people, it was the poison.

Several people were also killed with the 36th Battalion, particularly in the battle of the tunnel in Elisabethville. Most of the five or six who were killed in that battle came from A-company which was drawn from the eastern side of the country, including Dublin. We served close to them. We were only a few hundred yards apart. That morning, we took over several places including Camp Minière where the Force Publique was based. We cleared that out and one or two people were shot but didn't die. Katanga, where most of the wealth of the country was, had broken away from the rest of the Congo. Moïse Tshombe was the leader of Katanga. He was killed later, as was the Prime Minister of the Congo himself, Lumumba. The country was in fierce turmoil.

We guarded a lot of Belgians in a hotel in Manono. They were actually swimming in a pool that had clear drinking water, while we had no water to drink. We even tried to shave using orange juice and lemonade. You should try to get up a lather like that! The Belgians showed us very little gratitude for anything we had done. They were not very helpful. They never offered us a glass of water or something to eat, even though you could have been on duty for 24 hours. The Belgian people resented us to a great extent but they were glad to have us there to protect them. They resented the fact that we were there and they had to go. They had big businesses and were making a lot of money. We had to get them on planes and get them out safely.

When I was serving with the 38th Battalion, I met a few Belgians who were quite nice. We even had social contacts with them in a place called Kolwezi where we were guarding the airport. We would go to their houses and they would come to our officers' mess. Some of those Belgians were born and reared there. I can imagine that if you had a very good life and somebody came and took it away, you would be resentful. We did not take it away from them, but they were going and we remained there.

We were not prepared to cope immediately with the terrain, although we were able to handle the people. It seems to be a gift the Irish have for some reason. We can deal with people more easily than others can. We were not trained to use brute force and, given the weaponry we had, we would not have been able to do anything like that anyway. Mostly we had to protect ourselves and protect those who were put in our care. We did that quite adequately in the circumstances, as it was the first time we ever served abroad. The Irish were the babies going into uncharted territory because all the other UN forces had served somewhere overseas. Looking back on it, it does not seem to be half as bad as when you were there. At times we thought, 'This is it. It's over. We'll be overwhelmed', but it didn't seem to happen that way. At the time you coped with it, but when you think about it afterwards you say to yourself, 'Weren't we very lucky?' On numerous occasions you feared for your life. I remember at the farm in Elisabethville we had dug our trenches. They were firing mortars and two or three people were killed, including one lad who had only arrived that day. He had only been in a shack for a few minutes when a mortar killed him. He had only been in the Congo for an hour and a half. It seems that more men from the east of Ireland were killed than from the south, but that might have been because of their displacement. The 32nd Battalion went north and was not deployed in Katanga where most of the troubles occurred because of the secession. Some of our posts, in the 33rd Battalion, were more isolated; they were more in the bush than other companies, yet the others were unfortunate in that they travelled from a controlled environment into the bush on mobile patrols. That is where it happened. The men from Munster were fortunate enough, although a few were killed. We did not suffer as many losses as those from the Eastern Command.

We were able to return fire. If we saw targets that were firing at us we were permitted under the UN charter to defend ourselves and prevent ourselves from being forcibly disarmed. Other than that, you would always use diplomacy in as far as possible to resolve a problem, rather than dealing with it through firepower.

I saw the area at Niemba in which those people were killed, and it was a hostile environment with a wrecked bridge. It was a place in which I could imagine that, having been caught there, if I had been one of the people, it must have been traumatic, especially for the two lads that escaped. The memories they must have. It must be desperate altogether to see your comrades killed. They were caught in a way. They were trying to repair the bridge and some of them weren't armed. They were trying to build the bridge back up so they could cross over it. By modern standards the type of implements the Balubas had for killing people were ancient, but they were effective because of the numbers. Apparently there were hundreds and hundreds of these Balubas to our eleven men. Even if they were only throwing stones, they were going to kill you. However many Balubas you might have killed, they would have killed you anyway due to their overwhelming numbers.

I am sure that our involvement in the Congo served a purpose. Before we went out, I remember reading about some of the atrocities that were happening in the Congo and the fact that there was nobody there to do anything about it. Even if we had only been there as witnesses, it served a purpose. Before the UN troops were sent there, it was definitely a catastrophe because there was slaughter and no one was made to account for it. At least, as we were witnesses there, people had to account for their actions.

I did three tours of duty in the Congo and as the time progressed I saw more stability. When I went back with the 36th Battalion we were living in houses in Elisabethville. Our accommodation was reasonable and the cookhouses and administration had all been set up. People back home were sending stuff out. Guinness and the cigarette companies sent beer and cigarettes to us. We received food parcels that we had not had on previous tours of duty. We lived reasonably well in comparison to the first time I went there – you can imagine being dumped in the Congo with no communications and no back or forward lines.

When I went back with the 38th Battalion we were on the move all the time. We did five different towns or cities within the six months and we were looked after very well. At that stage the Belgian military had left, but there were still Belgian business people there. They remained with the consent of the local people because they ran the businesses and the locals were not capable of running such businesses themselves. The locals were getting more out of it, however, than when the Belgian army had been there.

JIM O'MALLEY, LIMERICK

I served twice in the Congo. My first trip was in 1960 with the 33rd Infantry Battalion. Fortunately enough for me I was not with A-company that was involved in the Niemba massacre. I served with B-company at Manono in the Albertville area. We had all assembled together for a few days before we departed. We got to know some of the men in A-company, not closely but just to see. I saw the scene of the ambush. It was quite a while afterwards before we went out to the actual spot. I did not see it at first hand.

There was a small, rough field airstrip at Manono. It was in the outback and far away from civilisation. We were not understood very well when we arrived there. We were white and it was shortly after Belgian rule so the locals took us for Belgians. Even though we were part of the UN force and wore blue helmets, they were still very suspicious of us. Being mistaken for Belgians was a fierce disadvantage because the only white people the locals had come into contact with were Belgians. As history shows, they were not treated very well by the Belgians. We heard stories of how they had been treated; they were working for half nothing and were practically slaves. There were copper mines at Manono and the Belgians had the locals working there for half nothing. I did not see any first hand evidence of brutality by the Belgians but stories were going around that there had been a bit of violence. The Congolese really hated the Belgians anyway and that should be enough evidence that something was amiss.

When the Belgians had been airlifted out we were in a difficult position. It took a considerable time – a month or two – to establish trust and to let the locals know that we were not Belgians, but we were there to establish peace and goodwill. We were there to help them rather than offend them. The day-to-day routine mainly involved patrols and camp security. In addition, we had guard duties in established areas including the copper mines, the stores area, the airstrip and our own encampment.

I did not see any fighting when I was there. The only hair-raising experience I came across was early on when they surrounded our camp. There was a certain amount of mistrust, as they did not know our true identity. We were treated with suspicion. They thought we were against them so the Balubas came with primitive weapons – spears and bows and arrows – and completely surrounded the camp for about ten hours. We were all dug in. Eventually the chaplain and the company commander got on to the Balubas' leader and the situation was defused. They returned to their villages in the surrounding areas and it was all quiet. We built up trust and got on okay. Although it was the same Baluba tribe, Niemba was 200 or 300 miles away from us. I would put the massacre at Niemba down to mistaken identity and the mistrust of white people at the time. It was an early period for us.

MICHAEL HIGGINS, CORK

I went to the Congo with the 32nd Battalion in July 1960. We in B-company were stationed in a place called Kindu, while the rest of the lads went to Goma. We used to go up into Samba where we trained in patrols. We met the Congolese on the retreat because there was a civil war going on. We took them all into our own camp and held them there until we could transport them out. Our toughest assignment was to wear the clothes we had. We went out with bull's wool uniforms, buttoned up to the neck and we wore those for approximately two months with the hobnail boots and gaiters. The food was very bad.

Tinned army rations were sent out from Ireland but they only lasted for a period because we were on our own most of the time. We had no cooks with us so the ordinary lads did the cooking. At the start they did not even know what the powdered potatoes were. Every time there was a meal parade you might get a case of peas so you would put out two or three cans of peas on each table. Somebody would open them and dish them out. They were not even heated half the time. We also got dog biscuits and bully beef. A lot of fellows lost a lot of weight, I can tell you, because the food was non-existent. I was a corporal at the time. We were living in the corridors of a hotel, not in the rooms. The officers lived in the rooms. The lads used to go on duty, marching up and down at the slope, the same way we used to do at home. Every ten minutes or less we would have to relieve them and send out another lot to do it because they just couldn't keep going. After two months we got a little bit of good clothing, so it was grand again. We were only two degrees away from the Equator at one stage. The temperature was always around 100 degrees Fahrenheit.

When I went back out again with the 36th Battalion, I came under fire. Our plane was hit while going in to land. Then we went up to headquarters at a place called the farm. There were two brothers called Fallon. One of them was killed by a mortar bomb inside the farm. The next morning we went out and took control of a police camp. We went from there to the tunnel in Elisabethville. Our C/O at the time was Bill Callanan. We were asked to retake the tunnel, which we did. We advanced at approximately 3.30 in the morning during a monsoon. We were in our positions and had taken the tunnel by about 7.30 that morning. There were no casualties in B-company. We went up the road. But there were a lot of casualties in A-company. They lost a lot of lads going up the railway line. The Katangese forces were firing on them, and the mercenaries most of all. That army had been made up of mercenaries and others, by Tshombe. Katanga was the richest part of the Congo because all the mines were there. They declared independence from the rest of the Congo. We were fighting them to take it back and hand it over to the government of the day.

When I arrived first, there were not many Belgians there. We had a good relationship with them. They were depending on us more than anything else and that's why there were good relations. We also had great relations with the coloured people. We were born to be peacekeepers. We took no sides. We went out as peacekeepers, nothing else. Believe it or not, a lot of them spoke very good English because they went to college in Ireland. Some of the local Congolese had gone to Galway, Dublin and Cork. You would swear to God you were talking to a Cork man half the time. They knew the history of the country as well because they had been educated by Irish priests. A lot of them were able to tell us about the division in our own country. We understood how their own divisions had come about. They trusted us.

When we went there with the 32nd Battalion we just didn't have a clue, not a clue. We didn't know what we were going into. Our dress alone would tell you that; we even had topcoats going out there. We were never informed what kind of a country it really was. The UN presence served a fantastic purpose there. We took Katanga province back. Everybody [in Katanga] was backing Tshombe who had declared himself president of Katanga. The rest of the country was poor. I thought it was great that we took this rich part of the country and gave it back to the Congolese who were the rightful owners. It was the right thing to do. It's like everything else, it was about greed. The Belgians were stuck behind Tshombe and the ordinary Congolese had nothing even though it was their country. It was handed over to them by the Belgians, yet they backed Tshombe when he led the breakaway province of Katanga.

By 1964 our job was done, but it would have been better if UN advisers, such as civil servants, had gone out there. They should have gone into government offices and shown them how to run the country. As we now know, their leader, Mobutu, made millions for himself; he made himself and his family rich but the ordinary person had nothing. The UN should have put somebody in to show them how to run the administrative side of the country.

We went to the Congo as peacekeepers with the 32nd Battalion, but we went out to war with the 36th and 38th Battalions, even though a lot people deny that fact. We fired at those people and they fired back at us, so it was war as far as I'm concerned. There were planes bombing us and the next time we brought in our own jets which were bombing them, so it was war. People may not agree that there was a war there, but there was. That's what it was all about. I am one of the proudest men in Ireland to have served in the Congo.

WILLIAM KEANE, LIMERICK

I served in the Congo with the 34th and 37th Battalions. Of all the operations that I have been on over the years, the Congo was the one place I always wanted to go back to, despite what went on there over the years. I always hankered to go back there if only for a holiday, although I know it is beyond the reach of most of us ever to be able to afford to go there. I have very good memories of my time there. We were so much younger in those days and the Congo was completely new. Overseas operations were completely alien to us. To be honest about it, we went there ill prepared as regards clothing and weaponry. To say the least we were primitive but looking back it was a much more adventurous operation than any of the others. It was not as formal. Every one of us down the line, from the top to the bottom, was in a learning process which made it that much easier.

I don't think we even realised the danger. Maybe the higher echelons did, perhaps they could see it better; but for us young lads, private soldiers and corporals, we didn't see the danger. I don't think any man ever looked back on it as being dangerous in that respect. I did not see any of my comrades killed or wounded. We were very fortunate not to have suffered any casualties but we did see some terrible atrocities that went on there against the civilian population and the various armed elements. There were quite a few of them there who were ruthless to one another, in the main. At times they hassled us but, personally, I wouldn't say that it was as stressful as the Lebanon or Cyprus. It is very hard to apportion blame to any one side for the atrocities, but you must remember that a civil war was going on there. It was soon after independence and the various factions were jockeying for position, so whoever was the strongest got the highest positions. Patrice Lumumba, who was Prime Minister of the Congo, was murdered at that time. Then Tshombe broke Katanga away from the rest of the Congo.

As regards atrocities, we were going into one particular village and there were bodies of natives actually strung up on the side of the road. All those who had been killed were male, needless to say, and they were just left hanging there, more as a deterrent I would say. Some of these atrocities were carried out by the white-led mercenary forces. It was their way of saying to the rest of the armed elements around the place, 'Keep out or this is what'll happen to you.' It was the first time that I had seen a dead body hanging on the side of the road. It is something we never thought would happen anywhere in the world. We were green but we learned fairly fast. I don't think that any soldier looks back on these things as being shocking; it is something that you learn to live with. After a while you get hardened to it. You come to a stage in life where you are not shocked any more. They had killed them, some of their bellies were opened and they just hung them up from the trees. It was like hanging up an election poster. It is a crude way of putting it but that's the way it was.

The Belgian population was not involved in that. The Belgians that remained there were very good to us, but we were probably their saviours. We were a buffer guaranteeing that they were going to get out. Those who stayed were members of the business community. A lot of those people, especially mining engineers, were necessary for the transitional period. Ourselves and the Swedes were the only white troops in that area because the other UN contingents were mostly made up of African or Asian nations. We had Pakistanis, Moroccans, Indians and Nigerians. There were other groups there also, but those four were the principal groupings. As far as UN troops are concerned, the colour of your skin doesn't really matter. It depends on the way you conduct yourself and your operation. You are not an occupying force, you are there purely as a police force. It is a very hard thing for a soldier under arms to be a policeman, but that is what we had to be. Our temperament suited the Congo. On a couple of occasions we went into villages and the natives would scatter because we had our bayonets drawn. We had one company commander, however, who decided to bring out the pipe band which was like a magnet. People came out of every hiding hole you could think of. Ever after that when we went into the villages we did so with the pipe band playing. Believe it or not, it had a great calming effect on the whole situation. As well as that, we shared nearly everything we had in the line of grub. It wasn't the Ritz by any means but we shared as much as we could, especially with the little kids. We got on marvellously with them and we made some good friendships.

Regardless of what went on there over the years, the Congo was the one place I hankered to go back to. I have very good memories of it. We were so much younger then, and it was new; overseas operations were completely alien to us. To be honest about it, we went there ill prepared in weaponry and ill equipped in clothing. To say the least, we were primitive but looking back it was much more of an adventurous operation than any of the others. It was not as formal. Every one of us down along the line, from the top to the bottom, were the UN forces, which made it that much easier. I don't think we ever realised the danger. Maybe the higher echelons did, perhaps they could see it better, but for us as young lads – Privates and Corporals – we did not see the danger. I don't think any man ever looked back on it as being dangerous in that respect.

We were very fortunate in that we did not suffer any casualties. We saw some terrible atrocities, however, that went on against the civilian population by the various armed elements there. Quite a few of them were ruthless to one another.

The UN forces undoubtedly served a useful purpose in the Congo, and on missions elsewhere. It is not for us as UN peacekeeping troops on the ground to say how good or how bad we are; it is how the general public perceives us after the operation. Once you go out there and do the best you can, and be as honest and fair as you can, you will succeed. It is very easy to go back and try to re-write history. If you want to re-write it in a political context you can do so in many ways. But the UN troops went in on the ground under the leadership of Lt General McKeown and it was a great honour for us as Irishmen to have an Irish supreme commander. It was a great achievement on his part to go in there and lead a multi-national force. You must remember that in those days the higher echelons of the Irish army didn't have that vast experience. But he went in purely as a soldier, not as a politician. It is very easy for politicians to come along and say this, that and the other – re-writing the whole thing. But when you are out there on the ground your job is to stop people from killing one another, basically, and to try to calm down the situation, to try to give a stable life to the native people. This goes for any country that the UN goes into. You are basically there to stabilise the situation and to try and stop the warring factions from killing one another. You do that to the best of your ability. When you go into any situation with a UN mandate, you are going in with a political mandate and you are trying to convert that mandate into a military solution. It isn't always the most ideal way of trying to do things but once you go in and try in all honesty to do the best you can, you can't go wrong.

31. Jimmy Clarke

JIMMY CLARKE, DUBLIN

I served with A-Company in the 36th Battalion. After some intensive training at the Glen of Imaal in County Wicklow we left by air for the Congo on 5 December 1961. We formed up at the infamous Camp Martini in Léopoldville, the capital of the Congo where we were paraded and informed of the situation ahead in Elisabethville, Katanga. Following a very quick meal we were informed that we were going into a war-like situation. The chaplain, Fr Cyril Crean (later parish priest in Donnybrook and since deceased) gave absolution to all concerned. We had a second chaplain, Fr Colm Matthews, who is now parish priest in Firhouse.

On 6 December, we flew out to Elisabethville to relieve the 35th Battalion. It was 1,000 miles further on from Léopoldville, just to give an idea of the extent of the Congo. It is a huge country. We received a very warm reception in Elisabethville to say the least. When one of the planes was coming in to land, one of its engines was knocked out by ground fire. When we eventually landed we found the end of the runway had been blown up, so our stores and ammunition had to be unloaded from the plane in double quick time to enable the pilot to fly back out. That was our baptism of fire. The airport had already been secured by other UN contingents. At its highest point the UN force was about 25,000 strong, including Indians, Pakistanis, Tunisians, Moroccans, Malayans, Danes, Swedish, Ethiopians and others, including ourselves.

We were then brought into Rousseau Farm, a UN military base in Elisabethville, from where we were conveyed to different locations. During that period the UN personnel came under attack from small arms and heavy weapons, day and night, by the Katangese gendarmerie loyal to Moise Tshombe, president of the breakaway province of Katanaga. Snipers were operating, too. Tshombe had set up his own government and the UN's task was to restore Katanga to the control of the central government in Léopoldville, which they did but with substantial loss of life.

A-Company had only been there two days when it suffered its first casualty, Corporal Mick Fallon, who was killed by mortar fire. Sergeant Paddy Mulcahy was injured and later died from his wounds. On 16 December 1961, during the battle of the tunnel in Elisabethville we lost two people in action – Lt Paddy Riordan and Private Andrew Wickham. In the days leading up to, and after, 16 December we suffered a number of casualties. Approximately 20 people were injured, some seriously and some not so seriously. As a driver, I was not in the front line but I was familiar with what was going on. There was intensive firing all around. I was armed with a Gustav sub-machinegun.

Operation Sarsfield, the attack on the tunnel, commenced at 0400 hours on 16 December. We had had no previous experience of this type of warfare. It was a new departure for all concerned. There was certainly an element of nervousness because we did not know where we were - under cover of darkness we did not know the lie of the land and didn't have sufficient time to become acclimatised to our surroundings. Most of the time we spent in trenches and conditions were very bad. It was the tropical rainy season so we were up to our waists in water in some of the trenches. The food was all tinned – bully beef, powdered eggs, milk and potatoes. That was it, there was no choice. You had two choices, take it or leave it! It was like something from the First World War. Our cooks were excellent, though. They turned out meals under dangerous and difficult conditions.

I was driving a pick-up truck. In the early days my job was to bring food and other supplies out to the trenches. I brought any casualties back to headquarters at Rousseau Farm for medical treatment. Prior to the battle for the tunnel, I brought back a number of casualties. One of them, Sgt Paddy Mulcahy, had been seriously injured and later died. The medical officers did not travel in the truck with me; they were based back at the medical aid post. I came under fire a number of times. They had snipers in the trees, so you did not know where they were.

On 16 December, the day of the battle for the tunnel, there was a torrential downpour. After many hours of fighting our men successfully took the tunnel, which was a railway bridge over a dual carriageway at the entrance to Elisabethville. Above the tunnel on the railway line, the Katangese forces were entrenched in train carriages. They had food and a substantial amount of ammunition. They were there for the long haul so we had to get them out, which we did. The battle ended after eight hours of fighting at around midday when the tunnel was taken. Some of the Katangese forces escaped, others surrendered and there were casualties also. There was a hospital to the right of the road, and a brewery on the left. Our troops did mopping up operations after that.

Following those hostilities, things slowly returned to normal. We held the area and eventually the UN commandeered a number of villas in the area, which had been abandoned when the Belgian population fled. The UN came to some agreement with the owners to rent them. Our company was housed in the villas, which were a lot better than the trenches we had been used to. The attack on the tunnel was a gradual build-up; you moved on, took an area and secured it, and then moved on again. Eventually, Operation Sarsfield came into play for the final push to take Elisabethville.

The commander was Comdt Joe Fitzpatrick who led the company into the attack. One of the platoon commanders, Lt Paddy Riordan, was killed in action and his radio man, Andy Wickham, was killed by the same burst of machine-gun fire. A number of others were wounded, and at least two had to be repatriated. In March 1962, we suffered our fifth loss – Corporal John Power, who died of natural causes in Elisabethville.

At the end of 1961, we were able to go into Elisabethville. It was a big, modern city with tree-lined boulevards, fairly big hotels and office blocks. The Belgians were still living there, running the city. They had all the businesses and accommodation. They were happy to get protection from the UN but at the same time they wanted to do their own thing, which they had been doing prior to then.

We did not realise it at the time, but the bigger political picture was all about the mines, which the Belgians, French and British companies controlled. They wanted their share of the wealth from the mines and we were ignorant of the political situation. In later years, we learned what it had all been about but at the time we did not know.

Eventually, people started being friendly towards us in Elisabethville. As things developed, we needed supplies and they needed money. The UN was prepared to spend money and we spent whatever money we had at the time, which was not an awful lot. We went into the city to buy things. We always wore our uniforms, we were not allowed to wear civilian clothes. I did not experience any animosity from the Katangese. There were no more hostilities after that, although there might have been a few scares but there were no further casualties.

When the Belgians there saw that the situation had returned to normal they gradually re-occupied their houses and re-opened their businesses. To a certain extent, there was tension between the Belgians and the local Congolese who had been suppressed. The Belgians were the masters who dictated the pace – who they employed and the working conditions. That was so, even during the breakaway period, which did not last long, and when everything got back to normal the conditions of the locals remained the same.

The local Katangese forces were not well trained but they were led by white mercenaries – French, British and German. They were the leaders and the locals took their instructions from them. There was an Irish mercenary out there called Major 'Mad' Mike Hoare, who was from Rush or Lusk in North County Dublin. I never met him personally. He was a mercenary leader – they were fearless and did not ask any questions. They went in to do a particular job – whether it was to get rid of some politician, cause upheaval or generate dissent – and got out afterwards.

Later on, we became more enlightened as to the big picture in the Congo, but we were not there for political purposes. It was mainly a military task and to the best of my knowledge we were the first UN troops to be ordered into battle. We went out there as peacekeepers but we became peace-enforcers when we were ordered into battle. That was the first time that had happened with UN troops. The Congo was the first foreign experience for the Irish Army. Prior to that, our troops had never been on overseas service, except as UN observers in the Lebanon and Sinai – they were small groups acting purely as observers.

In the early days in the Congo, Irish troops were stationed outside the big cities, in the bush. In fact, A-company of the 36th Battalion was originally bound for the bush. But sometime after our departure from Ireland a decision was made that we should not go to the bush but were instead to be deployed in Elisabethville because of Tshombe's secessionist action. We remained there for the duration of our service. The battalion consisted of: A-Company, traditionally the Dublin-based company; B-Company from the southern area, including Cork, Waterford, Kerry and Limerick; C-Company, traditionally from the western area, including Athlone, Mullingar and the Curragh command; and headquarters company, which was a combination of staff from various commands, including transport personnel, military police, medics, engineers and an armoured car section.

We finished up our six-month tour of duty in June 1962, and I returned to the Congo in 1963 to serve with UN headquarters in Léopoldville. Almost three years after independence, the city had still not been renamed Kinshasa. The second visit was a totally different experience because there was no conflict or hostility, although there might have been some internal, local stuff but we were not involved in it. Daily life was much more relaxed then. I got around quite a bit in Léopoldville, which was a huge, modern city with skyscraper hotels and outdoor swimming pools. Belgians and other Europeans were still living there. They remained on as long as they could but eventually they had to get out because their numbers were dwindling and the Congolese were gradually getting more control.

The UN mission ended in 1964 but inter-tribal warfare continued over the years. There was also an overflow of tribal populations into the Congo from Rwanda, Burundi and Angola. In 1994, the Hutu and Tutsis fought among themselves in Rwanda. The fighting in the Congo continues to this day, so I do not think the UN would have achieved any more than we did, by staying on after 1964. The sheer volume and numbers would have made it difficult. At the height of the trouble in Elisabethville, there was a refugee camp, which at one stage sheltered Balubas and others who had fled the bush into the city for protection. These tribes were rampaging all around the country, slaughtering people. They could not care less who they killed. We had a refugee situation to deal with, which was quite difficult because inter-tribal warfare was going on in the camp, but they were not allowed to have any arms there. Each tribe had its own splinter groups and protection rackets. The camp was full of women and children, the unfortunate innocent victims of the hostilities. At times we were

32. The river crossing near Niemba, scene of the Baluba attack in which eight Irish soldiers died on 8 November 1960.

stretched because we were responsible for the security of the refugee camp, plus other installations in the area. In addition, check points had to be manned and quite a lot of other daily tasks had to be performed. At one time there were about 60,000 refugees in the camp because they kept coming in from the bush. They fled the bush for protection because all the tribes were on the rampage. There were Balubas, Balubakats, Conakats, Pygmies, and Watutsis among others. They were all basically fighting for survival.

When I was in Elisabethville initially, there was some fear and apprehension on both sides – between us and the Congolese, including the Balubas. But they saw we were there to do a particular job and did it. When things returned to normal they were fed and given employment around the camps. They came to accept us then because they were not going to get it from any other quarter.

Some of our company's members had been out with the 33rd and 34th Battalions in 1960. It was the first time that two battalions had served together but that was done because the area of operations was so vast. It would have been impossible to control or police it with a single battalion.

We organised inter-company sports, including Gaelic football, hurling and soccer. We played soccer against UN teams from other countries. In March 1962, a group of entertainers were sent out from Ireland to entertain all UN contingents. They toured the Congo for a month under the direction of Harry Bailey. That was a great show and everyone looked forward to it because it was a link with home that we had not seen for some time. Our period of service included Christmas, and ran from December 1961 to May 1962. Our only contact with home was by letter, which was slow, but at least we had some contact. It would take a letter about ten days to reach home.

When we returned home it was great to get back to our families. Unfortunately, some people who went out did not come back with us. Over the years since then, we formed an association and we meet up every year to remember our fallen comrades. A-Company's annual commemoration consists of a wreath-laying ceremony in Glasnevin cemetery, followed by Mass in McKee Barracks. After that we have a get-together with tea, sandwiches and the usual refreshments. Most of our surviving members attend, along with the families of our deceased comrades.

A-Company, 36th Battalion, is the most decorated company in the history of the Defence Forces. As a result of many acts of bravery and courage, its members have been awarded fourteen distinguished service medals. It is a great achievement and one that is unlikely to be equalled. We take great pride in that fact.

Appendix I
Key Dates

1275 Kingdom of the Kongo founded by King Nimi.

1482 Portuguese explorer, Diego Cão, establishes a colony at mouth of Congo River.

1491 First Catholic missionaries arrive in Kongo and convert King Mani Kongo and his court to Christianity.

1498 Slave trade to America begins and will last for almost four centuries, until the abolition of slavery in the USA in 1865.

16th Century Establishment of the Luba Kingdom in central Shaba.

1641 Dutch take control of ports of San Thome and Saint-Paul.

17th century marks the start of slave trafficking from the southern Kasai and Katanga.

1815 Nabiembali establishes the Kingdom of Mangbetu (north-east Congo).

1816 Naval explorer, James Kingston Tuckey (born Cork, Ireland, 1776 – died Moanda, Congo, 1816) navigates the Congo River to the Yelala Falls, inaugurating the 19th century's scientific explorations in central and southern Africa.

1869 Chief Msiri founds the Kingdom of Garengaze in south-east Congo.

1874–8 Congo explored by Henry Morton Stanley (1841–1904) who navigates Congo river to the Atlantic Ocean.

1876 King Léopold II of Belgium sets up the International African Association with the goal of 'opening Africa up to civilization and abolishing the slave trade'. Léopold annexes the Congo as his own private property.

1878 King Léopold II strikes an agreement with Stanley to create settlements in the Congo and to negotiate treaties with local tribal chiefs.

1880 Catholic missionaries open first school in Congo, at Boma, catering for twenty children.

1883 Congo becomes the International Association of the Congo (AIC), presided over by King Léopold II (who never visited the country during his reign).

1884 At an international congress in Berlin, the AIC becomes the Independent State of the Congo, run by Léopold II with his government situated in Boma. Léopoldville (now Kinshasa) later became the capital city.

1885 Léopold II gains control of Katanga (now Shaba) by ceding other territories to France. Sir Francis de Winton is named as Administrator General to replace Stanley.

1888 Foundation of the Force Publique, a territorial army comprising African soldiers under the control of Belgian officers.

1889 Léopold II decrees that 'vacant land must be considered as belonging to the State.'

1891 Beginning of intensive exploitation of ivory and rubber for export. Force Publique defeats forces of Chief Msiri to establish state control in the south east. Chief Msiri murdered by a Belgian army officer.

1892 Geological expedition discovers vast mineral resources in Katanga. Belgians take control of the province.

1893–94 Belgian forces wrest control of eastern Congo from Arab traders.

1895 Force Publique troops mutiny in Luluabourg garrison. Ancient Kingdom of Mangbetu is conquered by state forces. A limited number of educated Congolese are registered as part of the 'civilised' population and are deemed entitled to the same civil rights as Europeans.

1896 Local population forced to grow cotton and cocoa.

1897 Railway line linking Matadi and Léopoldville opens.

1899 English novelist, Joseph Conrad, publishes his novella Heart of Darkness based on his horrifying experiences as a Congo riverboat pilot.

1903 Human rights abuses by King Léopold II's regime in Africa are denounced in Britain by the Congo Reform Association.

1904 International commission begins inquiry into brutal policies used to boost rubber production in the Congo, including amputations and hostage taking.

1906 Following uproar in parliament in Brussels, the Independent State of Congo is annexed by Belgium.

1908 The Congo becomes a Belgian colony and, until independence in 1960, will be run jointly by the colonial ministry in Brussels and a governor general in Léopoldville.

1911 Katanga is linked to South Africa by a new railroad.

1913 Start of industrial copper production. Diamond reserves discovered in the Kasai.

1917 Cotton cultivation begins in the Maniema region. Kasai diamond mines start operating.

1920 The Belgian minister for the colonies, Louis Franck, outlines a new policy for Africans in the Congo, including the establishment of chieftainships and indigenous tribunals. The Belgian Congo's first daily newspaper, *Avenir*, is published.

1921 Simon Kibangu founds a Christian group in the southern Congo. Kibanguism holds that Africans and Europeans are equal.

1924 International Red Cross opens an office at Pawa in northern Congo.

1925 The Kitawala Muslim sect develops at Maniema in central Congo and later in Katanga, seeking the swift departure of white settlers.

1928 BKC railway line opens, linking southern Congo and Katanga.

1931 Major revolt by the Bapende tribe in the north east, sparked by a fall in farm incomes.

1933 Two new provinces created (Kasai and Kivu) in addition to the existing four: Congo Kasai (later renamed Congo Léopoldville), Equateur, Orientale and Katanga.

1935 Maximum duration of labouring and crop production is set at sixty days per annum.

1939 Congo brought into World War II on side of Allies.

1941 First strikes by Congolese workers in Jadotville (now Likasi) and Elisabethville (now Lubumbashi). Strikers are massacred in Elisabethville.

1942 Limit of obligatory work extended to 120 days per annum.

1944 Workers' revolts in Katanga and Kasai, sparked by deterioration of miners' working conditions.

1946 Mutiny by Force Publique at Luluabourg.

1948 Belgian authorities recognize Congolese workers' right to strike, establish minimum wage and inaugurate public transport in Léopoldville.

1950 Establishment of Abako (Alliance of Bakongo), which develops into a major pro-independence party led by Joseph Kasavubu. Polygamy banned in the Belgian Congo.

1951 Establishment of the 'African Conscience' group.

1954 Opening of Louvanium University in Léopoldville.

1955 Royal visit by King Baudouin who launches the idea of a Belgo-Congolese community. Baudouin meets Patrice Lumumba (1925–1961), then active in trade unionism. In December, Belgian priest, Fr Antoine Van Bilsen (a professor at the Institute of Overseas Territories, Antwerp University) publishes '30-year plan' for granting Congo increased self-government. Van Bilsen's plan envisages the creation of a federal structure in the Congo and the training of a Congolese elite to progressively take over the 'levers of power'. The Belgian ministry of the colonies opens three colleges of administration in the Congo.

1956 The African Conscience association rejects Baudouin's plan for a Belgo-Congolese community but backs Van Bilsen's 30-year plan. Abako says 30-year plan is too long and demands immediate fundamental freedoms for Africans. Simon Kibangu's son, Joseph Diangienda, founds the Congolese national church. University opened at Elisabethville. Congo's first national football team, 'the Lions', formed (renamed 'the Leopards' in 1966).

1957 Balubakat formed by Jason Sendwe, as a Baluba umbrella group in Katanga. In December, local elections see sweeping gains for pro-independence parties, including Joseph Kasavubu's Abako faction which wins 130 of the 170 seats reserved for Africans. Kasavubu calls for general elections, press freedom and home rule for the Congo. Census puts native population at 13 million, with 108,000 whites of which 20,000 are settlers.

1958 Elected municipal councils are established on an experimental basis in Léopoldville, Elisabethville and Jadotville. Patrice Lumumba forms his MNC party (Congolese national movement) in Léopoldville. In October, Moïse Tshombe forms the Conakat group (confederation of tribal associations of Katanga). In December, Patrice Lumumba attends All African People's

Conference in Accra, capital of newly independent Ghana.

1959, 4–5 January: 49 Africans die (290 injured, including 50 Europeans) in violent suppression of an Abako pro-independence rally in Léopoldville. Formation of PSA (African Solidarity Party) by Antoine Gizenga.

11 January: Abako is banned by Belgian authorities.

13 January: In radio address, King Baudouin undertakes 'to lead the Congolese populations towards independence' with gradual introduction of self-government. He announces that local elections will take place on 20 December on a one-man, one-vote basis.

February–March: Riots continue in protest at arrest of Abako leaders. Kasavubu and other leaders released on 14 March but held in Brussels to prevent local demonstrations.

April: Conference of Congolese parties held in Léopoldville calls for establishment of Congo government in January 1961.

23 June: Abako leader, Joseph Kasavubu, seeks the establishment of a Republic of Congo in the west of the country.

30 October: Following riots in Stanleyville (now Kisangani) Patrice Lumumba arrested for inciting violence.

December: Martial law declared in southern Kasai to halt clashes between Lulua and Luba tribes.

16 December: King Baudouin visits Léopoldville. Pro-independence parties win 20 December election. Political leaders demand round-table conference on Congolese independence, to be held in Brussels.

1960, 29 January: Belgian delegates to the Brussels conference (which runs until 20 February) set 30 June 1960 as the date for Congolese independence. Period from end of round-table conference to independence marked by tribal unrest in Kasai, Katanga and elsewhere, while increasing numbers of Belgians leave the Congo.

10–18 May: Belgian parliament votes in favour of independence for Congo.

May: Following free elections in May, Patrice Lumumba's MNC emerges as largest single party winning 35 of the 137 seats. Joseph Kasavubu becomes president elect.

23 June: Lumumba is named as prime minister-elect.

30 June: Independence day ceremony is marred by clash between King Baudouin (who lauds King Léopold II's 'genius' and 'tenacious courage') and Lumumba (who attacks the Belgian Congo's 'regime of injustice, oppression and exploitation').

5 July: Force Publique troops mutiny, demanding pay rises and the sacking of Belgian officers.

7 July: Belgian troops intervene to protect Belgian civilians and put down the mutiny. New Congo government sacks Belgian general Janssens and appoints Victor Lundala (a sergeant-major) as army commander in chief.

11 July: With the support of Belgian business and 6,000 Belgian troops, Katanga declares its independence under the leadership of Moïse Tshombe,

leader of the local Conakat party (secession formally ends on 15 January 1963).

12 July: Lumumba and Kasavubu seek armed UN intervention to prevent civil war, and call on Belgium to withdraw her troops.

13–14 July: UN security council requests Belgium to withdraw its troops in favour of a 19,000-strong UN contingent.

27 July: Irish 32nd Infantry Battalion departs from Dublin for the Congo to join UN troops of 25 other nations there, under the banner of ONUC – the UN operation in the Congo. Lumumba appeals to USSR and Ghana for help in restoring order.

12 August: UN troops enter Katanga, as Belgian troops withdraw. Albert Kalondji, emperor of the Balubas, declares the diamond-rich South Kasai as an independent state (secession lasts until September 1962).

5 September: President Kasavubu sacks Lumumba (after only 67 days in power) and installs Joseph Ileo as Prime Minister.

8 September: US President Eisenhower appeals to USSR to cease unilateral interference in the Congo.

14 September: Colonel Joseph Mobutu seizes power in a military coup, suspending parliament and the constitution. Lumumba placed under house arrest, guarded by UN forces.

15 September: Irish UN troops called in to protect white civilians in Manono, Katanga, following Baluba attacks.

19 September: Mobutu installs a college of commissioners-general replacing parliament.

20 September: The Republic of the Congo is admitted to the United Nations as a sovereign state.

8 November: Eight Irish soldiers die in clashes with Baluba tribesmen at Niemba (Katanga). At least 25 Baluba die in same incident.

29 November: Lumumba escapes from house arrest and attempts to reach his supporters in Stanleyville.

2 December: Lumumba recaptured in Kasai province. Mobutu says he'll be tried for 'crimes against the state'.

1961, 1 January: Lt. General Seán McKeown appointed as commander of the UN's ONUC force in the Congo (until 29 March 1962).

17 January: Mobutu sends Lumumba by air from Léopoldville to Elisabethville airport where he is assassinated soon after his arrival.

9 February: A provisional government, led by Joseph Ileo, replaces the college of commissioners-general.

13 February: Official announcement of Lumumba's death.

21 February: UN Security Council authorizes UN troops to use 'all appropriate measures', 'if necessary, in the last resort', 'to prevent the occurrence of civil war in the Congo'.

7–12 March: Conference of Congolese leaders held at Tananarive (now Antananarivo) Madagascar (boycotted by pro-Lumumbist, Antoine Gizenga) agrees to form a confederation of Congolese states. Dr Conor Cruise O'Brien appointed as UN Secretary-General's special representative to Katanga.

26 April: Following all-party conference at Coquilhatville, Katanga's president, Moïse Tshombe, is arrested for condemning President Kasavubu's pact with UN to expel foreign militias from the Congo.

Mid-June: Dr Conor Cruise O'Brien arrives in Elisabethville, as UN Secretary-General's special representative, with mandate to end secession of Katanga.

2 August: Parliament votes to elect MNC co-founder Cyrille Adoula as prime minister, with Antoine Gizenga as deputy prime minister. UN troops begin disarming Katangese soldiers.

28 August: Operation Rumpunch (plan to neutralize the white leadership of the Katangan military) launched under direction of UN special representative, Dr Conor Cruise O'Brien. Irish, Swedish and Indian troops capture military posts throughout Katanga.

9 September: Tshombe's mercenaries seize control of Katangan gendarmerie to resist UN troops. Dr O'Brien launches Operation Morthor (Hindu word for 'smash'). In the confusion, Tshombe flees to Kolwezi.

11 September: Dr O'Brien granted international warrants for arrest of Tshombe and four of his ministers.

13 September: Tshombe flees across border to Ndola, Northern Rhodesia (now Zambia).

17 September: UN Secretary-General Dag Hammarskjöld dies in air crash on his way to Ndola to meet secretly with Tshombe in effort to negotiate a ceasefire. (In his memoir *My Life and Themes*, pp 230–4, O'Brien speculates that, in a botched hijacking, OAS men targeted Hammarskjöld 'whom they believed to have undermined the French empire in North Africa').

18 September: 155 Irish troops from the 35th Infantry Battalion taken prisoner at Jadotville, Katanga (after running out of food and ammunition).

20 September: Ceasefire pact with UN allows Tshombe to return to Elisabethville. Operation Morthor fails as secession of Katanga continues.

25 October: All 155 Irish troops released from custody in Jadotville and Kolwezi after five weeks' detention, following a prisoner-swap deal.

3 November: U Thant named as new UN Secretary-General. USA seeks new UN resolution authorizing 'use of any and all force necessary' to overcome the mercenaries in Katanga.

2 December: Conor Cruise O'Brien resigns his UN post. (According to his memoir, *My Life and Themes*, pp 242-58, O'Brien felt he was the victim of a British plot to oust him and, thus, foil the UN bid to end Katanga's secession).

On 16 December, Irish troops take part in the 'battle of the tunnel' for control of Elisabethville.

1962 Stand-off continues between UN and Katangese forces. UN launches Operation Grand Slam, a full-scale assault on Katanga's political and military infrastructure. Secession of Southern Kasai ends. On 28 December, UN troops take over key positions in Elisabethville. Widespread fighting ensues between UN troops and Tshombe's forces.

1963, 15 January: UN forces take full control of Elisabethville, ending the secession of Katanga.

September: President Kasavubu suspends parliament, as opposition parties go underground.

December: Pierre Mulele unleashes a revolutionary war in Kwilu.

1964 Forces loyal to Gaston Soumialot seize the east of the country and establish a breakaway government in Stanleyville. Meanwhile, Christophe Gbenye's Lumumbists seize control of Kivu and Orientale provinces.

March: Belgium halves debt owed to Brussels by Congo.

May: End of Irish involvement in Congo when 2nd Infantry Group withdraws.

June: End of four-year UN mission to the Congo.

10 July: Moïse Tshombe is named prime minister of new national government with task of ending regional revolts.

4 August: Stanleyville falls to rebels without a fight.

November: Belgian, British and US troops are parachuted into Stanleyville to end the secession there.

1965 23 October: President Kasavubu appoints Evariste Kimba to replace Tshombe as prime minister.

25 November: Coup d'état by General Mobutu topples Kasavubu and Kimba. Political parties banned. The Mobutu dictatorship would last 32 years.

1970 Mobutu becomes president.

1971 Congo renamed Zaire.

1973–4 Mobutu nationalizes many foreign-owned companies and forces European investors to leave.

1977 Mobutu invites foreign investors back. French, Belgian and Moroccan troops repel attacks on Katanga by Angolan-based rebels.

1989 Zaire defaults on loans from Belgium. Development programmes cancelled as economy deteriorates.

1990 Mobutu ends ban on multi-party politics but retains substantial powers.

1991 When unpaid soldiers riot in Kinshasa, Mobutu agrees to formation of a coalition government.

1993 Rival pro- and anti-Mobutu governments formed.

1994 Mobutu appoints Kengo Wa Dondo as prime minister. One million people die in Rwandan genocide, as Hutus attack Tutsi tribe.

1996 Tutsi rebels capture much of eastern Zaire while Mobutu is abroad for cancer treatment.

1997 Mobutu flees the capital in May 1997, ahead of Rwandan-backed rebel troops loyal to Laurent Desiré Kabila, who takes over as the new president. Zaire renamed Democratic Republic of the Congo.

7 September: Mobutu dies (of prostate cancer) in exile in Morocco.

1998 In August, rebel forces, backed by troops from Rwanda and Uganda, advance on the capital, Kinshasa. They are repulsed following intervention of forces from Zimbabwe, Namibia and Angola. In October, rebels capture the government stronghold of Kindu in the east.

1999 Anti-government rebels hold half territory of Congo. In August, all rebel groups sign up to peace accord in Lusaka, Zambia.
2000 In February, UN Security Council authorizes a 5,500-strong force to monitor ceasefire. Ethnic fighting erupts in the rebel-held eastern Congo.
On 2 May, Brussels parliament opens inquiry into possible Belgian government involvement in death of Patrice Lumumba in 1961.
2001 16 January: President Laurent Kabila shot dead by one of his bodyguards in the presidential palace in Kinshasa. His 31-year-old son, Joseph Kabila, takes over as president. On 23 November, a Belgian parliamentary inquiry finds that Belgian government ministers bore 'moral responsibility' for the events leading to the murder of the Congolese independence leader Patrice Lumumba in 1961.
2002 Belguim apologizes for its role in the murder of Lumumba.

33. UN Secretary-General, Dag Hammarskjöld, arrives at Léopoldville Airport, 1961. On the left, Lt General Mobutu and President Kasavubu. On the right, Irish General Seán McKeown, head of UN forces in the Congo January 1961 to March 1962

Appendix II
Record of Irish Army's Unit Service
with ONUC
(United Nations Mission to the Congo)
July 1960 to May 1964

Unit	Number of Soldiers	Period of Service
32nd Infantry Battalion	689	July 1960 – January 1961
33rd Infantry Battalion	706	August 1960 – January 1961
34th Infantry Battalion	648	January 1961 – June 1961
1st Infantry Group	340	May 1961 – November 1961
35th Infantry Battalion	654	June 1961 – December 1961
36th Infantry Battalion	715	December 1961 – May 1962
37th Infantry Battalion	723	May 1962 – November 1962
2nd Armoured Unit	96	October 1962 – April 1963
38th Infantry Battalion	730	November 1962 – May 1963
3rd Armoured Unit	89	April 1963 – October 1963
39th Infantry Battalion	464	April 1963 – October 1963
2nd Infantry Group	337	November 1963 – May 1964

The total number of Irish soldiers who served in ONUC, the UN Mission to the Congo, was 6,191 (of which 501 were Officers, 1,808 NCOs and 3,882 Privates).

Sources: Irish Military Archives
        UN Training School Ireland.

Appendix III
Baluba Report on Niemba Battle,
8 November 1960 (translation)

Territory of Manono
Kiambi Post
                                        Kiambi, 15 November 1960

Re: Kasanga chieftainship's war against the UN.

I, the undersigned, Louis Mambwe, commandant of the Kiambi gendarmes, present the following report:

When we left Kiambi on 9 November 1960, Chief Kasanga and his president, André Ngoie*, told us that they had fought that battle the day before – that is to say, on 8 November (*note:– he is the President of the Kasanga Senate, imposed by the Niemba administration to the aforementioned chief).

This is how it happened:

1. We were told that those against whom we had fought were UN people. That is not possible. If they had been UN personnel, they would not have been patrolling with some of our enemies among them. This is the third time they have come to our area, because we don't have a war here. We only want to fight the Kiambi people; those are the ones whom we seek. We have learned that they are real men and that's why we want to fight them.
2. Those Europeans, from the Niemba post, came here for the third time. What was their motive? It's 35 km from Niemba to Kasanga, and at the 25-km mark there runs a small river called the Lweyeye; we destroyed the bridge spanning it because we feared our enemies would come that way to enter our territory. We are all members of the coalition here, as are the Kiambi people. Although the Europeans constantly repaired the bridge, we demolished their work each time. They asked us if the Kiambi people had come to demolish their work on the bridge and we replied, 'Yes'. Then they told us they wanted to fight against the Kiambi people. The first time they said that, they went back because it was impossible to advance.
3. Another day, they came to repair the bridge again and said they were from the UN. We let them do it but we thought that if they were from the UN, they could not fight against the coalition. They returned to Niemba saying they'd come back in three days.

4. Then we asked ourselves how we would know if they were from the UN or were our enemies, the next time they'd arrive at our barrier.

5. Our commander – an ex-sergeant major, first class – said we should dress in our coalition clothing, including leopard-skin headdress, as if preparing for combat. The UN people had already seen coalition warriors dressed like this. If they were really UN people, they would not do anything.

6. When it was time for them to go to Kiambi, our commander called his gendarmes to enact the battle. He positioned his advance guard at the barrier to our territory. Then he told us that when these men came, he would approach them dressed as a member of the coalition. If, at that moment, they fired their rifles, it would mean they weren't UN personnel but our enemies.

7. In fact, when they arrived on the other side of the Lweyeye, our commander approached them. He put a leopard-skin on his head, in the manner of the coalition [forces] and left with four gendarmes.

8. The commander went in front of their truck, with his gendarmes, and pointed to the leopard-skin on his head. All of a sudden, without warning, he was struck by bullets, as were the four gendarmes. The commander fell dead on the spot.

9. Seeing that their commander had been shot dead, along with the four others who had followed him, the remaining gendarmes, who were armed, unleashed a hail of arrows.

10. When these enemies saw the coalition forces firing arrows, they increased the firing rate of their own weapons, which included a machinegun and a Bren gun. Having seen this, the coalition forces understood that these people did not intend to flee and were behaving like enemies. They [the coalition forces] continued to fire their arrows.

11. In truth, many enemy and coalition personnel died there. The battle began at 3 o'clock and finished at 5 o'clock. Bodies were scattered everywhere.

12. Now, we are accused of having waged war against the UN, whereas it was they who were at fault: (a) it was they who started shooting; (b) they fired non-stop for two hours, until the end of the battle – why?; (c) we saw no difference between the UN and our enemies because they didn't stop firing.

13. If they were from the UN, they would have stopped firing for a second or a minute, and the battle would not have continued. In any case, there would not have been so many gendarmes killed.

Kasanga, 9 November 1960
Commandant Louis MAMBWE

Source: Prof. Daniel Despas, Belgium
Editor's Note: The word 'Cartel' in the original French text refers to Balubakat Cartel. The abbreviation Balubakat stands for the Association of Balubas in Katanga.

Appendix IV
(Original French text)

Kiambi, le 15 novembre 1960

OBJET: guerre de la Chefferie Kasanga contre l'ONU

Je, soussigné, Mambwe Louis, Commandant des Gendarmes de Kiarnbi présente le rapport suivant:

Nous avons quitté Kiambi le 9 novembre 1960, le Chef Kasanga et son Président, Ngoie André, nous ont dit qu'ils avaient fait cette bataille la veille, c'est-à-dire le 8 novembre. (note.- il s'agit du Président du Sénat de Kasanga, imposé par l'Administration de Niemba au Chef précité). Cela s'est passé ainsi:

1. On nous a dit que ceux contre lesquels nous nous sommes battus sont des gens de l'ONU. Cela nest pas possible. S'il s'était agit des gens de l'ONU, ils ne se seraient pas promenés avec certains de nos ennemis parmi eux. C'est la troisième fois qu'ils viennent ici chez nous, car nous n'avons pas de bataille chez nous. Nous voulons seulement nous battre avec les gens de Kiambi. Ce sont ceux-là que nous recherchons. Nous avons appris que ce sont de vrais hommes, et c'est pour cela que nous voulons nous battre avec eux.

2. Ces Européens au poste de Niemba viennent ici pour la troisième fois. Pour quel motif? Entre Niemba et Kasanga il y a 35 km et au km 25 se trouve un ruisseau qui se nomme le Lweyeye. Nous avons démoli le pont qui l'enjambe parce que nous craignons que nos ennemis ne passent par là pour entrer dans notre pays. Nous autres tous ici, sommes des membres du Cartel, tout comme les gens de Kiambi. Or ces Européens réparent constamment ce pont, alors que nous démolissons leur travail à chaque fois. Ils nous demandent alors si ce sont les gens de Kiambi qui sont venus démolir leur ouvrage. Et nous répondons que oui. Ils nous disent alors qu'ils veulent se battre contre les gens de KIAMBI. La première fois qu'ils ont dit ca, ils sont retournés car il n'y avait pas possibilité d'avancer.

3. Un autre jour, ils sont venus à nouveau réparer le pont, disant qu'ils étaient de l'ONU. Nous les avons laissé faire, mais nous avons pensé que s'ils étaient de l'ONU, ils ne pourraient se battre contre le Cartel. Ils sont retournés à Niemba disant qu'ils reviendraient dans trois jours.

4. Nous nous sommes alors demandé comment nous ferions savoir s'ils étaient de l'ONU ou s'ils étaient de nos ennemis, lorsqu'ils se présen-

teraient ici à notre barrière.

5. Notre Commandant, un ancien premier Sergent Major, dit que nous devrions mimer un combat, que nous devions revêtir notre tenue du Cartel avec des peaux sur la tête. Les gens de l'ONU ont déjà vu des guerriers du Cartel en cette tenue. Si ce sont vraiment des gens de l'ONU, ils ne feront rien.

6. Leur jour d'aller à Kiambi étant arrivé, notre Commandant appela ses Gendarmes pour faire son simulacre de bataille. Il installa son avant garde à la barrière de notre Pays. Puis il nous dit qu'au moment où ces hommes arriveraient, il irait au devant d'eux, habillé comme un membre du Cartel. Si, à ce moment, on tire des coups de fusil, c'est que ce sont non pas des Onusiens, mais bien nos ennemis.

7. En fait, lorsqu'ils arrivèrent de l'autre côté de la Lweyeye, notre Commandant alla au devant d'eux. Il se posa une peau sur la tête, à la manière du Cartel, et parti avec quatre gendarmes.

8. Le Commandant étant arrivé devant leur camion, avec ses gendarmes, indiqua la peau qu'il avait sur la tête. Tout aussitôt, sans rien demander, il fut frapper par les balles, ainsi que ses 4 gendarmes. Il tomba et mourut sur le champs.

9. Là -dessus les gendarmes qui étaient armés, voyant que leur commandant était tombé sous les balles ainsi que ceux qui le suivaient, lancèrent une grêle de flèches.

10. Ces ennemis voyant que les Cartels lançaient des flèches, augmentèrent le tir de leurs armes. Ils avaient une mitrailleuse et un fusil mitrailleur. Ce que voyant, les Cartels comprirent que c'était des gens qui n'avaient pas envie de s'enfuir, et qu'ils se comportaient en ennemis. Ils continuèrant à tirer leurs flèches.

11. Vraiment, il mourut là -bas, beaucoup d'ennemis et de Cartels. La bataille commenca à trois heures et se termina à 5 heures. Il y eu des cadavres épars.

12. Maintenant, on nous accuse d'avoir fait la guerre contre l'ONU. Cependant, ce sont eux qui sont en faute. Ce sont eux qui ont commencé à tirer. Ils ont laissé tirer pendant deux heures sans arrêter jusqu'à, la fin de la bataille, et cela pour quoi? Nous n'avons vu aucune différence entre l'ONU et nos ennemis parce que nous n'avons pas vu qu'ils ont arrêter de tirer.

13. S'ils avaient été des Onusiens, ils auraient arrêté de tirer pendant une seconde ou une minute, et la bataille n'aurait pas continué. En tous cas, il n'y aurait pas eu tant de gendarmes tués.

Kasanga le 9 novembre 1960
LE COMMANDANT
MAMBWE Louis

Editor's Note: Grammatical mistakes have been left unchanged as they appear in the original French-language text.

Appendix V

Extract from 33rd Battalion's Unit History (pp. 101–2) on Discovery of Remains of Trooper Anthony Browne in 1962

The story of the NIEMBA ambush was not yet ended.   TPR. BROWNE'S body had not yet been found.   Information which we had received from Baluba survivors in MANONO hospital led us to believe that he had been killed immediately after saving PTE. KENNY, and that his body had been removed from the scene by the ambushers.

In the Autumn of 1962 COMDT. P.J. LIDDY, at that time Legal Officer, 37 Infantry Battalion, visited the Palais de Justice in ELIZABETHVILLE.   In conversation with a Belgian advocate, M. de BRUYN, who had been concerned in the trial of the NIEMBA murderers, he learned that the location of BROWNE'S body was known.   After numerous enquiries in early November a team of four officers was sent to the NIEMBA area with a marked map to locate the body.   They were COMDT. S. GALLAGHER, 37 Infantry Battalion, COMDT. M. McMAHON, ex Legal Officer, 33 Infantry Battalion, who was sent out from IRELAND specially, COMDT. B. HEANEY and CAPT. LAVERY both ex A Company, 33 Infantry Battalion, who were now in the advance party of 38 Infantry Battalion.   With an escort of Malayan U.N. troops and a small party of A.N.C., the area was at this time under the control of the Central Government,   they arrived at NIEMBA on Tuesday 6 November.   After an anxious night surrounded by openly hostile villagers who denied all knowledge of an ambush or of the body of a white soldier, they set off next morning for the village of TUNDULA.   After a stormy conference with the NIEMBA civil administration and the village elders, during which they were surrounded by large numbers of trigger-happy young warriors with arrows at the ready, the search began. After some journeying through the bush between the main road and the river the Administrator led them to the spot where TPR. BROWNE'S body lay.   The date was 7 November '62, two years almost to the day from the ambush. Apparently some days after the ambush, wounded, exhausted and starving he had called some women at the outskirts of the village and asked them for food and directions to the railway line, offering them 200 francs.   They took his money but instead of helping him they told the young men of the village who came out and killed him. A miserable death for the young hero who by his bravery and selfless conduct at the bridge a few days before had earned the first Bonn Calmachta to be awarded in our Army.   His remains were placed in a box and taken to ELIZABETHVILLE.   From there they were flown home to Ireland for a soldier's funeral and a Christian burial beside his comrades in the Congo Plot in GLASNEVIN.

Source: Private

Appendix VI
Report – Recovery of Remains of Tpr Browne in the Niemba Area (November 1962)

<u>REPORT – RECOVERY OF REMAINS OF TPR BROWNE IN THE NIEMBA</u>

AREA 5/7 NOV 62

1. On Mon 5 Nov 62, four IRISH officers - Comdts T.M. McMahon, B. Heaney, J.F. Gallagher and Capt Lavery left ELIZABETHVILLE by air for the purpose of recovering the remains of Tpr Browne, who, it was believed, was killed at the time of the ambush at NIEMBA.

2. The group arrived in ALBERTVILLE and remained there overnight. At 0900 hrs Tue 6 Nov 62, a patrol, consisting of four IRISH officers, a MALAYAN escort of one officer and 25 OR, an adjutant and 3 OR of ANC and an interpreter, left by road for NIEMBA.

3. At 1530 hrs the patrol reached NIEMBA where it encountered an ANC garrison of platoon strength. We had been informed that the ANC had made arrangements for liaison but the garrison seemed to know nothing of our mission.

4. At 1600 hrs, we reached the area of the ambush position. In accordance with information at our disposal, we called at TONDULA, a village which lies SOUTH fo the NIEMBA position. We tried to contact the headman of the village but failed to do so as he had fled to the bush. It was decided, then, to search an area near the village where, it was thought, the remains might be. This had just started when the local Administrator arrived together with some ANC officers. The Administrator's name was MUSAMBA LAMARIHE. He was rather excited and pointed out that he had had NO notice of our arrival. On learning what our task was, he grew even more excited and pointed out that in arriving without warning we had frightened the natives into the bush. Furthermore, he stated that they would now stay away and he would find it difficult to get their co-operation.

5. We assured him that the purpose of the patrol was to recover the remains and that we were NOT interested in anything else. He was also told that we intended to stay until the remains were found. He agreed to talk to the villagers and promised co-operation but persisted in saying that the natives were afraid. This was understandable when it is considered that the local inhabitants had almost certainly participated in the 'NIEMBA ambush'. The effect on the people of the sudden arrival of a UN patrol looking for the remains must have been electrifying. At the time, our impression was that the Administrator himself was quite frightened.

6. The patrol stayed at NIEMBA for the night. Little sleep was possible in the circumstances. The ANC garrison was billeted in nearby houses and it was felt that they would side with the natives against UN troops. The state of fright of the natives was such that they might resort to aggressive action and they had at least the night in which to lay plans. The reaction of the whole area was difficult to assess accurately.

7. Since we regarded the situation as rather tense, we examined the security measures we would have to take on the following day. It was realised that any search of an area would disperse the patrol and make it more liable to ambush. It was decided -

    a. that half of the patrol would remain on guard in the trucks at all times during the search;

    b. that at NO time would the local officials be allowed to slip away from the patrol;

    c. that the IRISH officers would at NO time be isolated from the remainder of the patrol.

A.

8. At 0830 hrs 7 Nov 62, the patrol met the Administrator and moved off to the ambush area. On the way, we encountered the local chief, Chef KASANGO de ét a NIEMBA and with him we went to a village which is the nearest village to the bridge where the ambush took place and is about 1,500 yds from the bridge. A long initial parley took place between the Administrator and the elders of the village. After about 45 minutes, the IRISH officers were invited to the conference. It seemed that there was considerable disagreement among the elders about whether the remains should or should NOT be given back. This was NOT stated as such but was camouflaged by a discussion on whether or NOT they would help in the search for the remains.

9. The local Administrator seemed to have made up his mind that the natives should co-operate, but there was disagreement among the village elders about it. A vehement argument took place and this lasted for about an hour. In fact, we were unable to contribute much to this phase. We felt that the Administrator was on our side and that he would be helped by our leaving it to him. We thought it sufficient to keep calm, and to reassure the natives that NO punishment was to be meted out.

10. During this time, it was difficult to assess what was going on literally behind our backs. The MALAYAN patrol could NOT deploy into a full scale security position because the natives were most suspicious and we were afraid that it would jeopardize the success of our mission. The patrol stayed out of sight on their trucks but, nevertheless, near enough to interfere if necessary. On one occasion, a truck-load of MALAYANS did pass the village and immediately the natives showed concern.

11. Finally, the elders agreed to help in the search. We were certain that they knew exactly where the remains were but it was necessary to carry out the form of a search.

12. The search commenced with a line of searchers leaving the road and searching towards the river. The IRISH officers, the Administrator and guides were on the right of this line. Almost immediately after the search began, the guides veered sharply to the right and moved at quite a sharp pace through the bush. Immediately, the problem arose of keeping the patrol together. It was felt that to slow the guides down unduly might jeopardise the operation. Apart from requesting them to go slowly, we did nothing else. The result was that the pace was most uncomfortable.

13. Capt LAVERY was in close contact with the guides. I was 20 yds – 50 yds behind him. Comdt McMAHON followed me and Comdt HEANEY was further behind. Several times Comdt McMAHON called out to me to slow down the pace as it was much too fast. I, in turn, called out to Capt LAVERY but I did NOT insist on the guides stopping because I thought that if the guides were sternly halted they would 'panic', and might well disappear into the bush and we would NOT see them again. There was the risk that we were being led into an ambush but again although it concerned us greatly, we felt that there really was NO alternative but to take the risk.

14. We travelled about 1½ miles through the bush at a very fast pace, when suddenly the guides stopped and motioned to us. Capt LAVERY and I went about 30 yds in the direction indicated and found the remains lying in an area of about 20 yds in diameter. Within five minutes, Comdt McMAHON joined us. He complained about the pace of the search, but I said that we really had had little initiative in the matter and I had felt that to interfere too much with the guides would have frightened them off.

15. When the remains had been gathered, we returned to the village. It was now 1200 hrs. Heavy rain began falling so we decided to leave for ALBERTVILLE.

16. The road from NIEMBA to ALBERTVILLE is for the most part a narrow dirt track. It runs across the 'grain' of the terrain and one has to cross a succession of ridges. In between ridges are deep ravines, along the bottom of which streams flow. The bridges across these streams are flimsy and the approaches to the bridge on each side are very steep. After the heavy rain, the red dust became mud and great difficulty was experienced in negotiating the approaches to bridges and the bridges themselves. The patrol had two Bedford trucks with field tyres, and several times the trucks slid broadside-on down the steep slope towards the bridge. The MALAYAN drivers xxxxxxxxxxxxxxxxxxx in each case were first-rate and I have NO doubt that a less proficient driver would have come to grief. On one occasion, a truck did slide broadside on into a large rut in the side of the road and it took some time to get it out.

17. The patrol reached ALBERTVILLE at approximately 2000 hrs on 7 Nov. 62. Comdt McMAHON and myself returned to ELIZABETH with the remains of Tpr BROWNE on 8 Nov 62. Comdt HEANEY and Capt LAVERY returned on 9 Nov.

Conclusion

18. Since none of the IRISH officers spoke FRENCH OR SWAHILI, we were compelled to rely on our TANGANYIKAN interpreter (JOHN MAKOKHO). At the end of the patrol, we were able to interrogate him in full with regard to the conversations to which he had listened and in which he had been involved.

19. His reports were interesting and were as follows:-

    a. The villagers found it difficult to believe that the expedition was NOT punitive.

    b. The local Administrator was extremely frightened at the whole prospect of searching for the remains. However, once he realised the situation, he talked the villagers into assisting the patrol. It is likely that his reputation has been enhanced in the area as a result of the whole affair. He became very cocky towards the end and led the interpreter to believe that he knew exactly what had happened at the ambush and subsequently. He did describe details of the ambush.

    c. The adjutant of the ANC who had accompanied us from ALBERTVILLE had privately advised non-co-operation with UN and had stated that UN were enemies of their country.

    d. During the 7 Nov, a group of young armed natives had been hidden in the bush near the conference village, ready to attack the patrol if it made any aggressive move.

20. There was NO way of corroborating these reports of the interpreter, except to say that in all other respects, he was efficient and accurate. Several times during the meeting, he had reflected accurately the attitude displayed and had enabled us to assess the atmosphere reasonably accurately.

John F. Gallagher

John F. Gallagher Comdt

An Coláise Míleata.

Source: Private

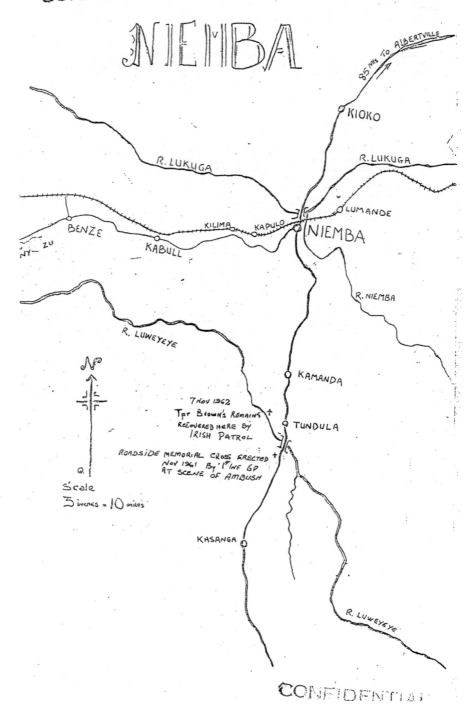

Notes

INTRODUCTION

1. In June 1958, the Irish Army sent 50 officers as unarmed observers attached to UNOGIL (the UN Observer Group in Lebanon). The observer group was wound up in December 1958. In addition, from December 1958, Irish Army officers have been sent as unarmed observers to serve with UNTSO (the UN Truce Supervision Organisation). That mission is still continuing and to date some 285 officers have served with UNTSO in Egypt, Lebanon, Syria, Israel, Iran and Iraq.

CHAPTER 3

2. *Memoir: My Life and Themes* (Dublin: Poolbeg Press, 1998, p. 205).

CHAPTER 9

3. Brigadier-General Hogan died in retirement in Cyprus, on 11 March 2004, aged 84.

CHAPTER 10

4. Thought to be Colonel Thomas Gray, the Army's Director of Operations at the time.
5. The 33rd Infantry Battalion's unit history (pp. 91–3) contains an unsigned three-page document, purporting to be a statement by Private Thomas Kenny on the events at Niemba.

CHAPTER 11

6. Conakat: Confederation of Tribal Associations of Katanga.

Index